IMAGES
of America

PICATINNY ARSENAL

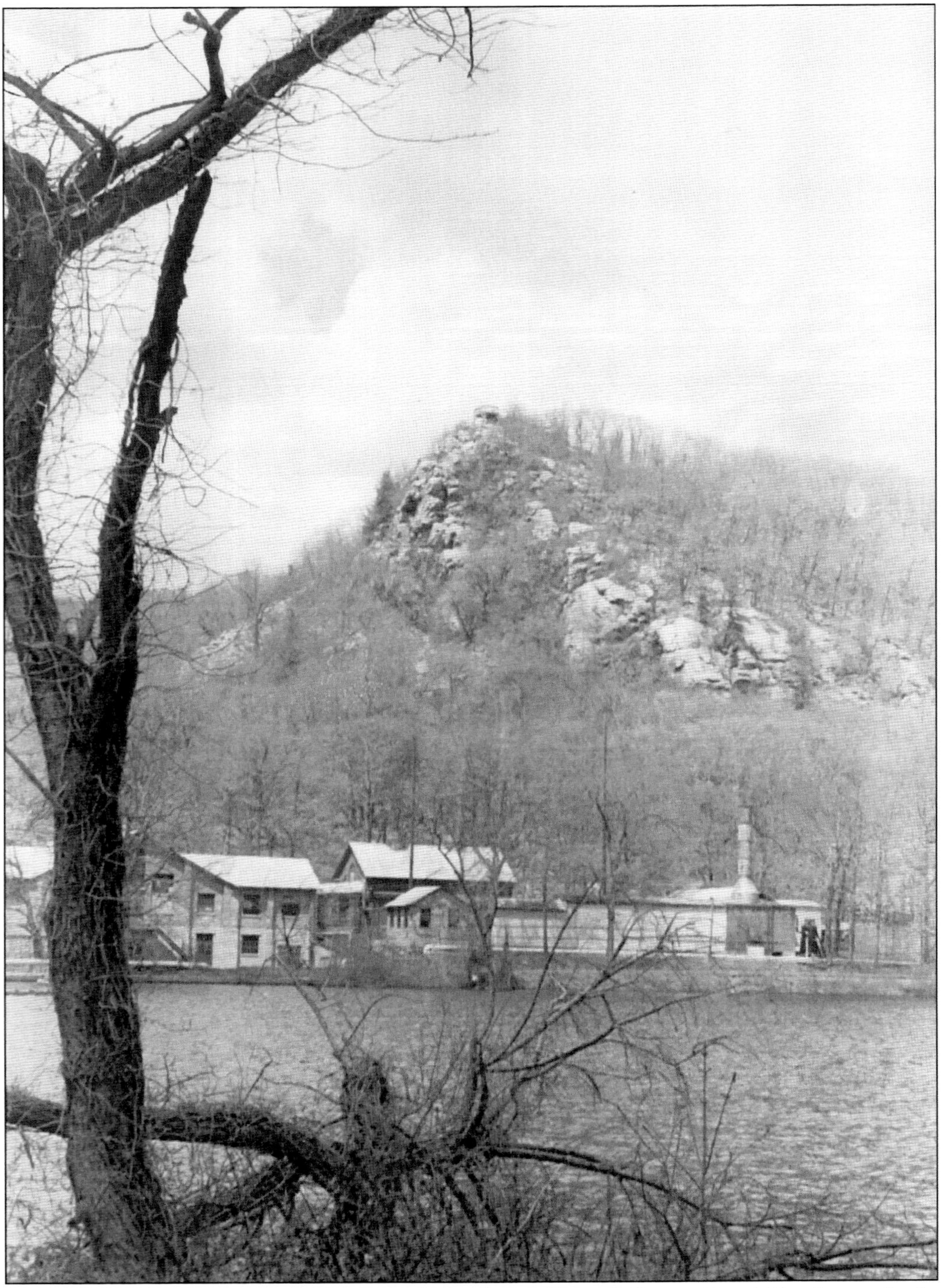

PICATINNY PEAK. A mountain rising from the valley of Green Pond Brook overlooks Picatinny Lake, a 109.5-acre artificial lake, and the arsenal. In 1749, Johnathan Osborne purchased land, built a dam, and erected a forge. It produced "cannon, shot, bar iron, shovels, axes, and iron implements for Gen. George Washington's Continental Army." (U.S. Army Photograph-Picatinny Arsenal.)

IMAGES
of America

PICATINNY ARSENAL

John W. Rae

ARCADIA
PUBLISHING

ISBN 978-0-7385-0196-3

Published by Arcadia Publishing
Charleston, South Carolina

Printed in the United States of America

Library of Congress Catalog Card Number: 99063932

For all general information contact Arcadia Publishing at:
Telephone 843-853-2070
Fax 843-853-0044
E-mail sales@arcadiapublishing.com
For customer service and orders:
Toll-Free 1-888-313-2665

Visit us on the Internet at www.arcadiapublishing.com

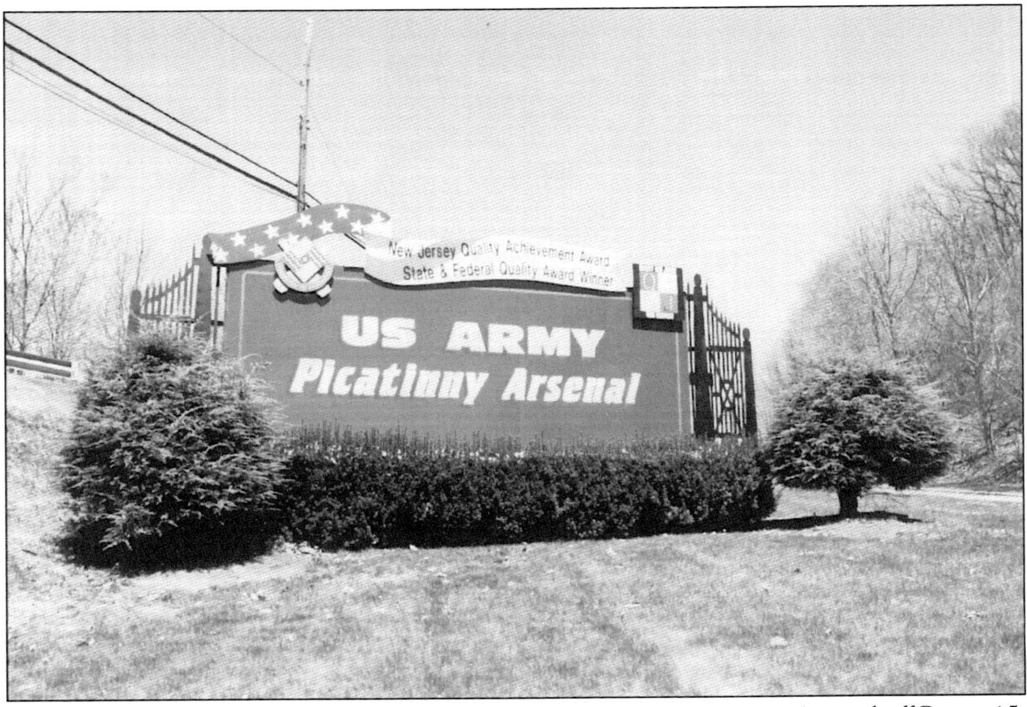

PICATINNY SIGN. This sign is located at the main entrance to Picatinny Arsenal off Route 15, 5 miles north of Dover, New Jersey. Erected in the early 1980s, it is the only indication that the 6,500-acre arsenal lies in the distant valley. On either side of the sign are wooden replicas of the Cannon Gate, the main entrance until the purchase of additional land in 1941. (Author's Archives.)

Contents

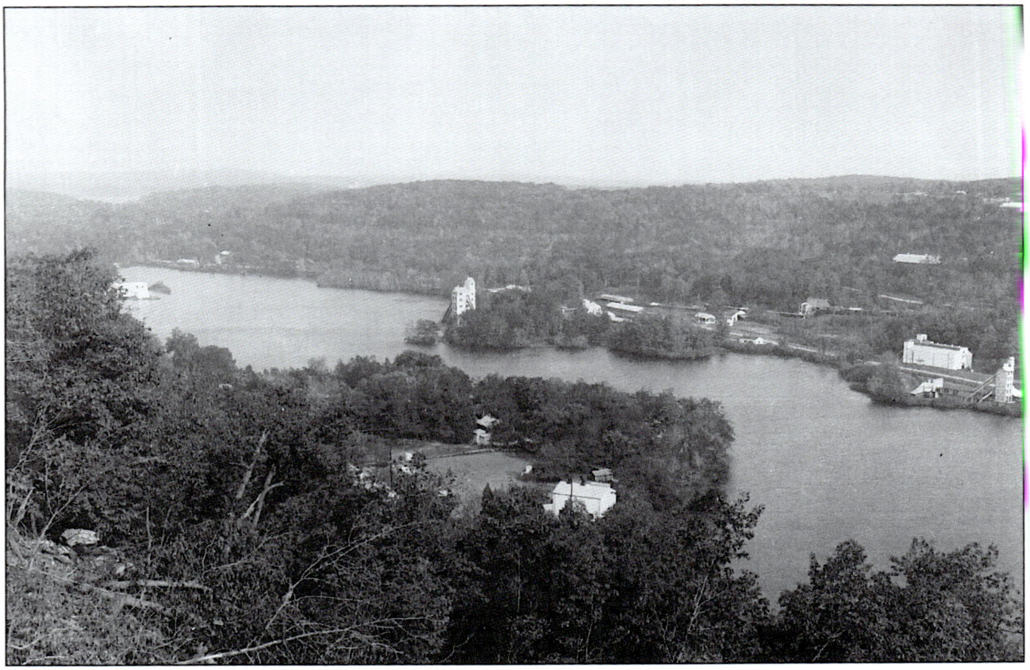

AERIAL VIEW. Shown is the upper end of Picatinny Lake, a 109.5-acre artificial body of water, and the main buildings, laboratories, magazines, and warehouses surrounding it. In the foreground is the melt loading area; on the opposite shore are powder blending buildings, and on the left a magazine. (U.S. Army Photograph-Picatinny Arsenal.)

ACKNOWLEDGMENTS

This photographic history, the first published on Picatinny Arsenal and its role as the cradle of American ammunition research, development, and manufacture, traces the 6,500-acre site from pre-Revolutionary War days through World Wars I and II, the Southeast Asian conflicts, and Desert Storm. Made possible by the assistance of Congressman Rodney P. Frelinghuysen, R, 11 Dist. NJ, it is not the work of one person, but of a long list of personages and historical societies.

Without the assistance of Brig. Gen. John P. Geis, commanding general (TACOM-ARDEC), and Patrick Owens, arsenal historian and museum curator, who answered an endless list of questions and made available the arsenal photographic archives, it could not have been compiled. It is sad, but so much that can be revealed has already been lost: pictures and negatives, the identity of persons in images, buildings that were burned together with their production records, and dates of important events.

The story of the arsenal is evident in its museum. Although, it suffers from lack of funds to maintain its endless list of accomplishments, exhibits, and weapons and to preserve the tanks, shells, rockets, containers, anti-aircraft guns, and rockets in dire need of restoration and paint in the museum park.

John W. Rae

INTRODUCTION

Near Lake Denmark in Morris County, New Jersey, a peak called Picatinny rises from the valley floor. In 1907, the United States Powder Depot located nearby became Picatinny Arsenal. The sprawling 6,500-acre Arsenal remains a powerhouse of great minds, both civilian and military, that work to ensure the U.S. Army is at its best on the battlefield.

As early as the Revolutionary War, forges located at the foot of Picatinny Peak and Lake Denmark made everything from cannonballs to shovels and axes for Gen. George Washington's Continental Army. Thus began a historic partnership with America's military and a Picatinny contribution to almost every major battle our nation has fought, including America's most recent missions in the Persian Gulf and the Balkans.

In 1879, Picatinny, known then as the U.S. Powder Depot, provided storage space for large quantities of pyrotechnics—powder and explosives. During the Civil War, the need became apparent for such a depot, and the future Picatinny Arsenal rose to the occasion. For more than a decade, Picatinny remained merely a storage and powder depot. Then, a few years before the Spanish-American War, Picatinny began to assemble powder charges for cannon. In 1902, Picatinny became a storage arsenal for reserves of sodium nitrate and armor-piercing projectiles; four years later, in 1906, the original powder factory was erected at a cost of $165,000. A year later, the U.S. Powder Depot became Picatinny Arsenal.

For a time during this period, in addition to the Army, there was a U.S. Navy presence at Picatinny as well. In 1891, the Army transferred to the Department of the Navy 315 acres at Lake Denmark. Over the course of the 69 years the Navy was at Picatinny, it was designated as the U.S. Naval Powder Depot, where shells for the coastal defense of New York Harbor were stored. In its last role, which ended in 1960, it was the U.S. Navy Rocket Test Station.

Perhaps the Second World War reinforced Picatinny's vital role in our nation's military. When the Japanese bombed Pearl harbor on December 7, 1941, Picatinny Arsenal was the only military installation in the nation capable of producing anything larger than small arms ammunition. While private industry quickly began converting to war production, Picatinny's work force practically expanded overnight to almost 20,000 men and women.

It was no accident that Picatinny was ready to meet the challenge of playing a major role in arming U.S. soldiers at the start of World War II. While the United States has no military ammunition industry in peace time, Congress has always provided essential funding for the Army for research, development, and engineering work—the primary mission of Picatinny, even today.

At the outbreak of World War II, Picatinny helped set up loading plants throughout the country, provided training for key defense personnel from the U.S., Australia, and Canada, and aided the nation in the procurement of equipment for new munitions plants. As Hitler encroached on lower European countries, most of the pilot operations at Picatinny expanded to all-out assembly line production. Without the industrial expertise developed at Picatinny, the rapid conversion of commercial entities to mass munitions manufacturing would have been impossible. Throughout the war, Picatinny trained 8,000 men and women in the techniques of mass-producing munitions ranging from artillery shells and bombs to rockets. Amazingly, while this production and training were going on, the great minds of Picatinny were simultaneously designing, testing, producing, and shipping—with incredible speed, I might add—new weapons for America's war overseas.

Following World War II, Picatinny again filled the gap until private industry could re-gear itself to resume the role of producing bombs and other standard munitions as the nation confronted wars in Asia—from Korea to Vietnam.

In 1950, when the North Korean Army started using Russian-made T-34 tanks, American soldiers discovered that the Bazooka shells they were using just bounced off the tanks. Gen. Douglas MacArthur issued a call for larger shells, and Picatinny was ready to meet the challenge. A test plant producing larger shells expanded to full-scale production in less than 24 hours. As these rocket shells were finished, they were trucked to nearby Morristown Airport where waiting Army cargo aircraft were on standby to fly them across the Pacific. Days later, North Korean tanks were destroyed.

Today, Picatinny remains the nation's major development arsenal for research and engineering of military propellants, explosives, and hardware for army weapons of all kinds. A highly skilled civilian work force has always been the hallmark of Picatinny Arsenal.

Over the years, Picatinny's dedication to advancing our military's rate of safety and success is unparalleled. In recent years, Picatinny has led the Army in its ability to use "smart" weapons in battle. Wars of the modern day are fought with modern technology, computers, and electronics. The great minds of Picatinny continue to be on the cutting edge of developing what the Army refers to as the weapons of the Army After Next—the army of the 21st century. All the while the men and women who work at Picatinny continue to win some of the Army's highest honors and awards for their mission to provide a quality environment, excellent facilities and services.

In New Jersey, Picatinny Arsenal is an important part of our community. And for our nation Picatinny's research and development mission is critical to the success of America's defense forces. Right now, on any given day, there are about 165,000 members of the U.S. Army on duty in 70 countries. As I write this, America's troops are actively engaged in the conflict in the Balkans, and an on-going mission in Iraq. Each one of our men and women in uniform holds a piece of Picatinny's technology in his or her hands, wherever our troops are stationed around the world.

While as a nation, the sense of duty and self-sacrifice of our men and women in uniform serves as an inspiration for us all. In New Jersey, we are just as proud of Picatinny Arsenal, and the vital, historic role it continues to play in keeping our troops safe and successful.

Rodney Frelinghuysen
Member of Congress
Washington
May 14, 1999

One

The Formative Years
1749–1925

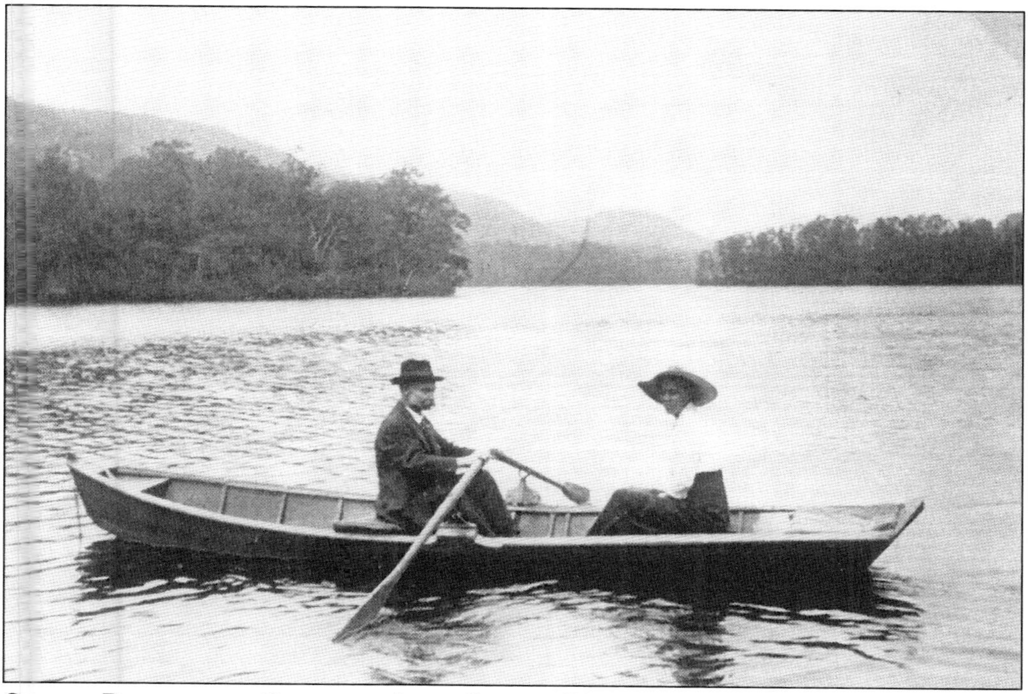

Couple Boating on Picatinny Lake Before World War I. The arsenal maintained a boathouse on the shores of the lake and rowboats for recreational use. The lake contains 2.5 million gallons of water, 3,000 of which flow over the spillway per minute. (U.S. Army Photograph-Picatinny Arsenal.)

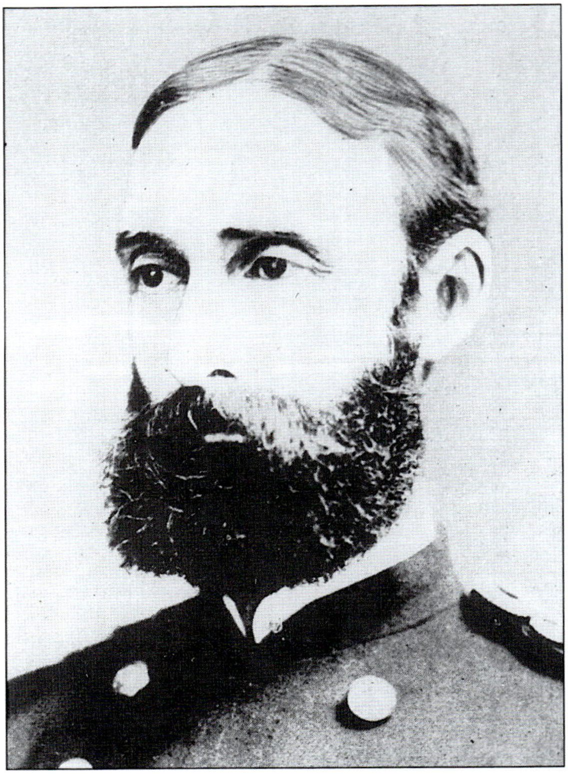

MAJ. F.M. PARKER, FIRST COMMANDING OFFICER OF THE PICATINNY POWDER DEPOT. Immediately after approval by Congress March 16, 1880, Parker started negotiations to purchase the initial 1,866-acres of land at a cost of $62,750. In addition, he obtained a 50-foot wide strip of land one mile long from the depot to the Dover-Sparta Road for $200. (U.S. Army Photograph-Picatinny Arsenal.)

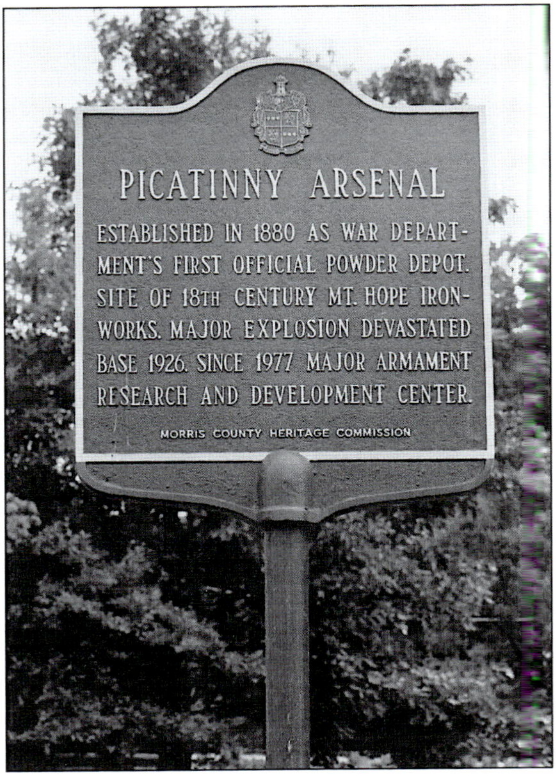

HERITAGE COMMISSION SIGN. The sign erected near the main entrance gate by the Morris County Heritage Commission in June 1985, outlines three major phases of the arsenal's development as a powder depot, the 1926 explosion, and as the army's research and development center. (U.S. Army Photograph-Picatinny Arsenal.)

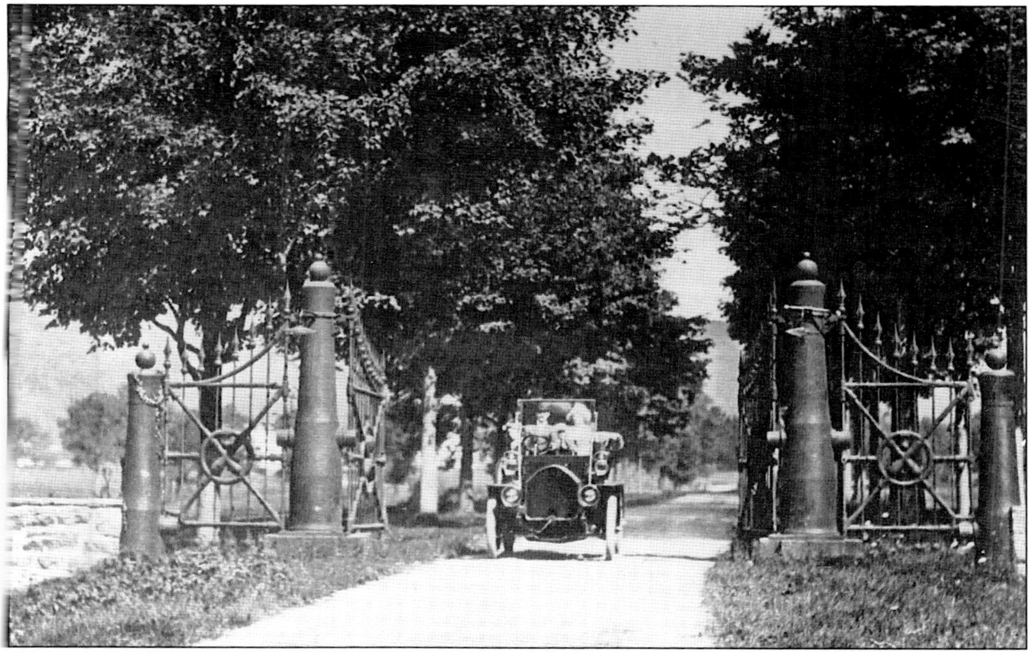

CANNON GATES. A couple in a vintage car approach the Cannon Gate in 1907 *en route* to the Dover-Sparta Road (Rt. 15), one mile distant over the one lane dirt road (sometimes called Government Road). The gate marked the arsenal boundary until 1941, when additional land was purchased and a three-lane entrance road built. (U.S. Army Photograph-Picatinny Arsenal.)

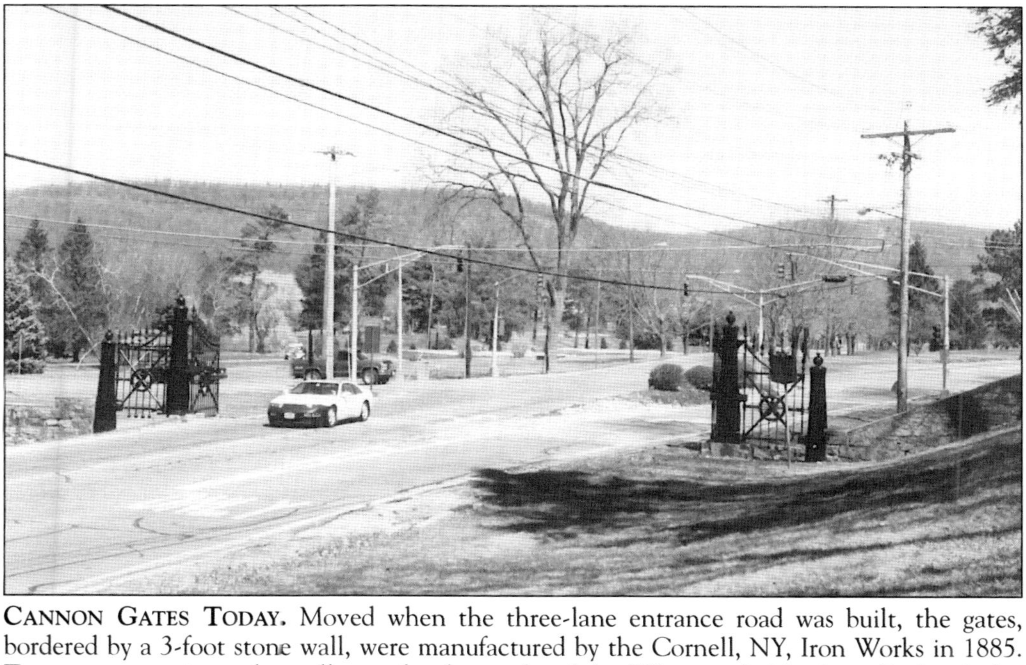

CANNON GATES TODAY. Moved when the three-lane entrance road was built, the gates, bordered by a 3-foot stone wall, were manufactured by the Cornell, NY, Iron Works in 1885. Two entrances pierce the wall, one for the tracks of the Wharton & Northern Railroad, the other for a road paralleling the tracks. The gate insignia is the badge of the Ordnance Corps. (Author's Archives.)

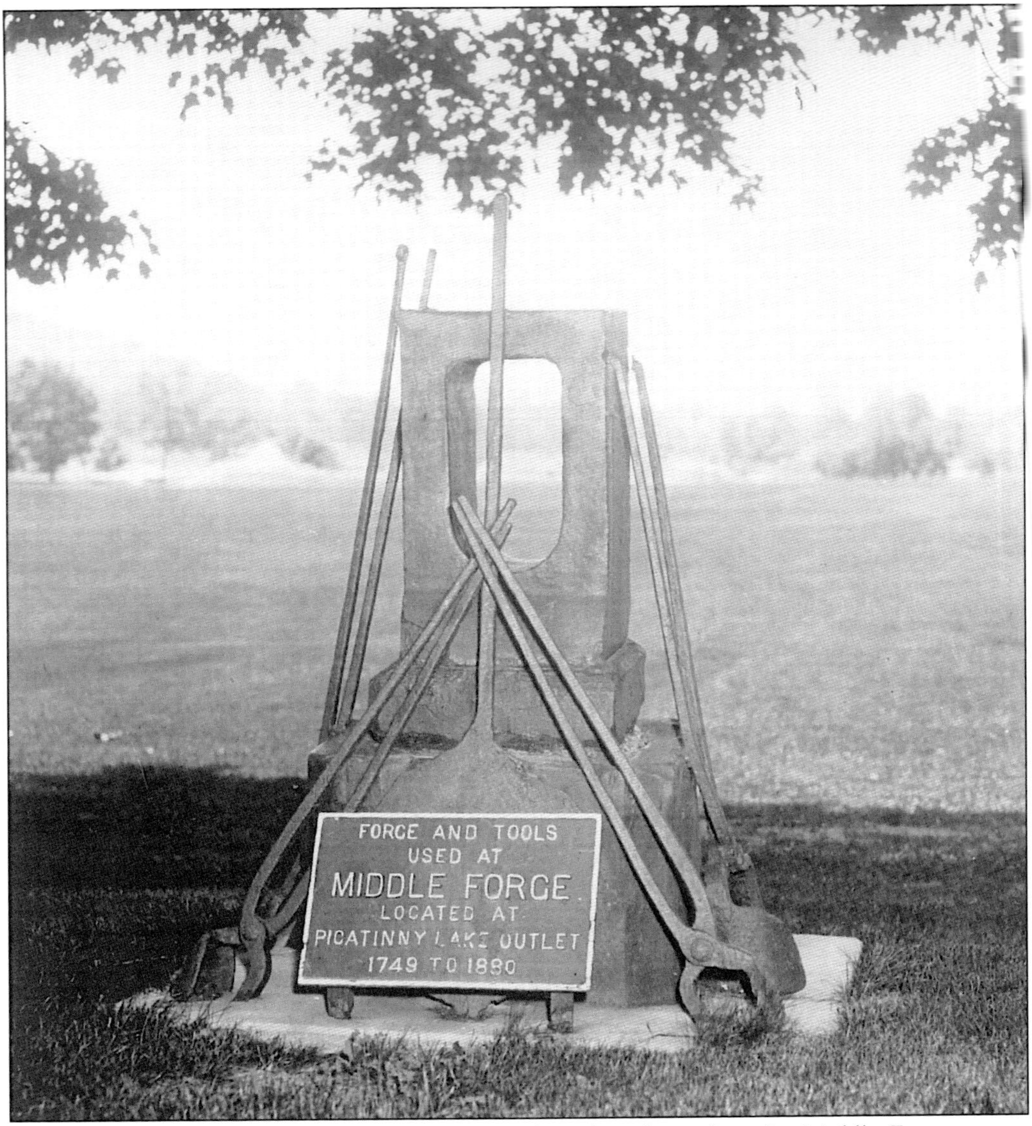

MIDDLE FORGE TOOLS. The trip hammer, anvil, and tools used at the Middle Forge are on exhibition at the arsenal. Dating to 1749, they were used by 250 Hessian prisoners who cut wood, burned charcoal, and operated the forge in the Revolutionary War using iron from nearby mines. The forge was sold to the government in 1880. (U.S. Army Photograph-Picatinny Arsenal.)

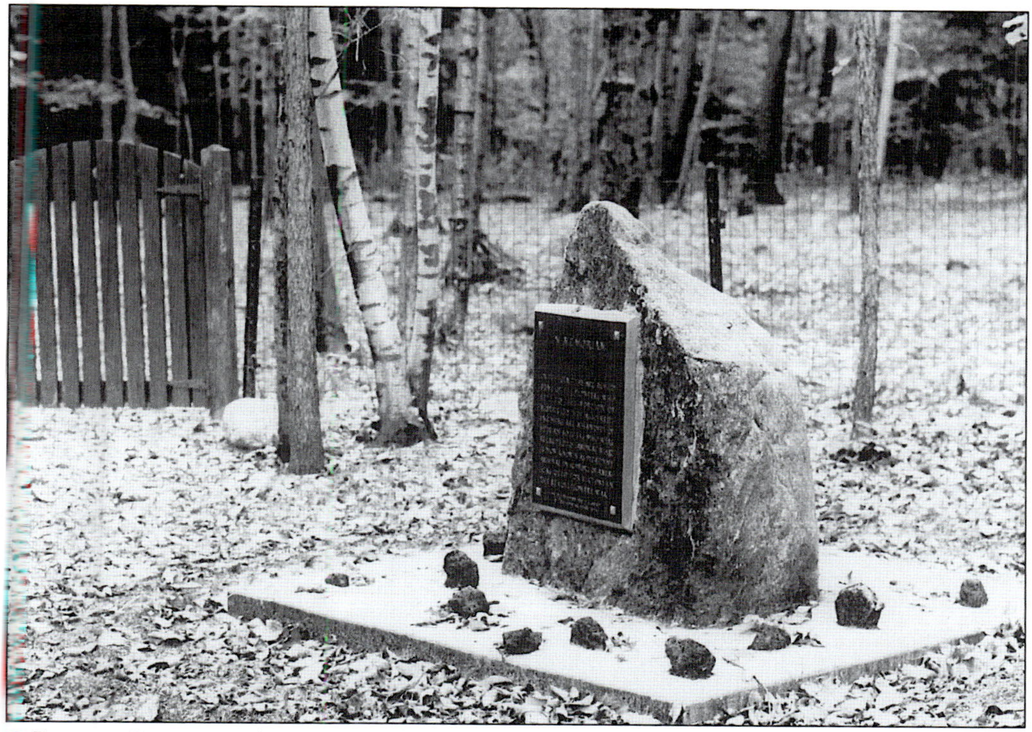

A BURIAL GROUND. This stone marker is located at the Walton Burial Ground, also called the Hessian Cemetery, due to unproven belief that Hessian prisoners who worked the forge may be buried there. Only a few headstones, their markings erased by time and weather, remain. (U.S. Army Photograph-Picatinny Arsenal.)

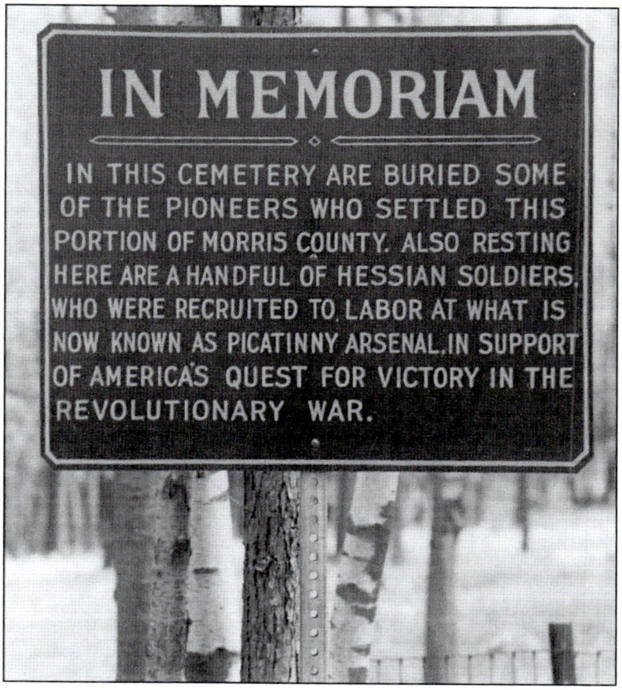

IN MEMORIAM. This sign at the cemetery states a handful of Hessian prisoners sent to assist local workers in the mines and forges by Gen. George Washington rest in the cemetery beside early settlers of the region. (U.S. Army Photograph-Picatinny Arsenal.)

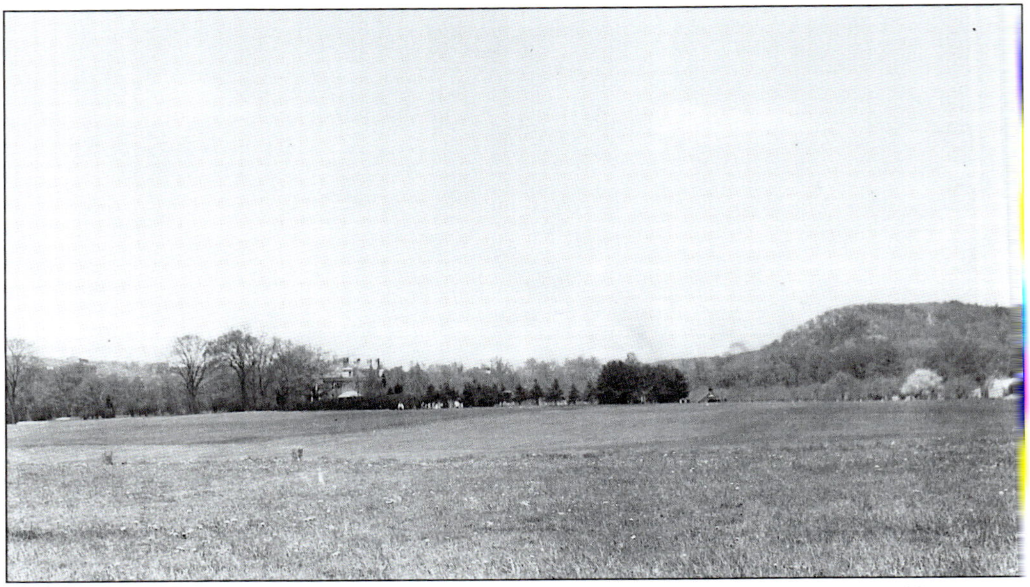

FIRST AIR STRIP. The initial landing field was constructed on what is now the golf course in front of the commanding officer's residence, visible in the background. Used until World War II, it was 2,800 feet long. A modern helicopter field of the New Jersey Air National Guard has superceded it. (U.S. Army Photograph-Picatinny Arsenal.)

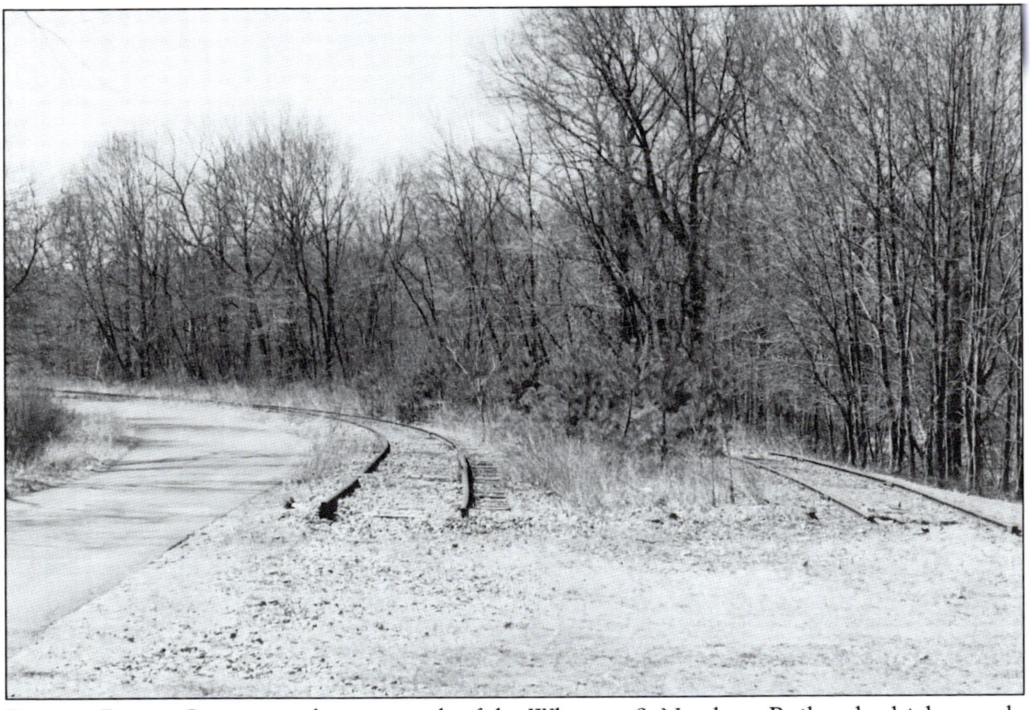

RUSTED RAILS. Grass grows between rails of the Wharton & Northern Railroad, which as early as 1887 operated a passenger service to the arsenal for employees. Early timetables show stops at Picatinny Arsenal, The Factory, Navy Depot, and Lake Denmark. With the exception of these tracks at the Cannon Gate, most tracks were ripped up in 1979. (Author's Archives.)

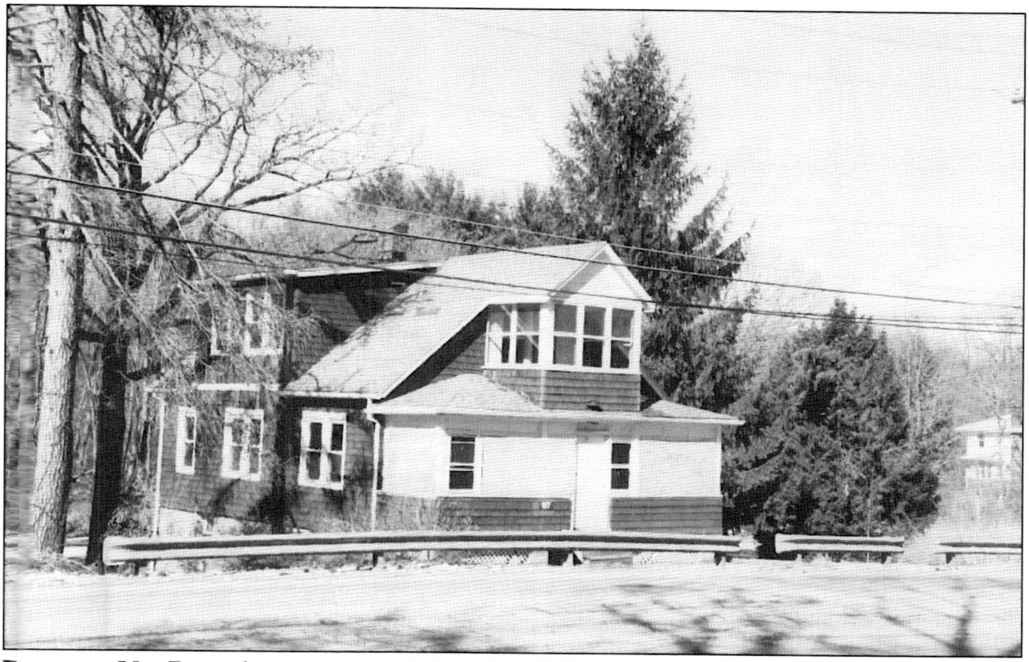

BOARDED UP. Frame houses, many of them boarded up and built shortly after the turn of the century in what was then Spicertown, line Parker Road, the main entrance to the arsenal. They served for years as quarters for married officers' families stationed at the arsenal. (Author's Archives.)

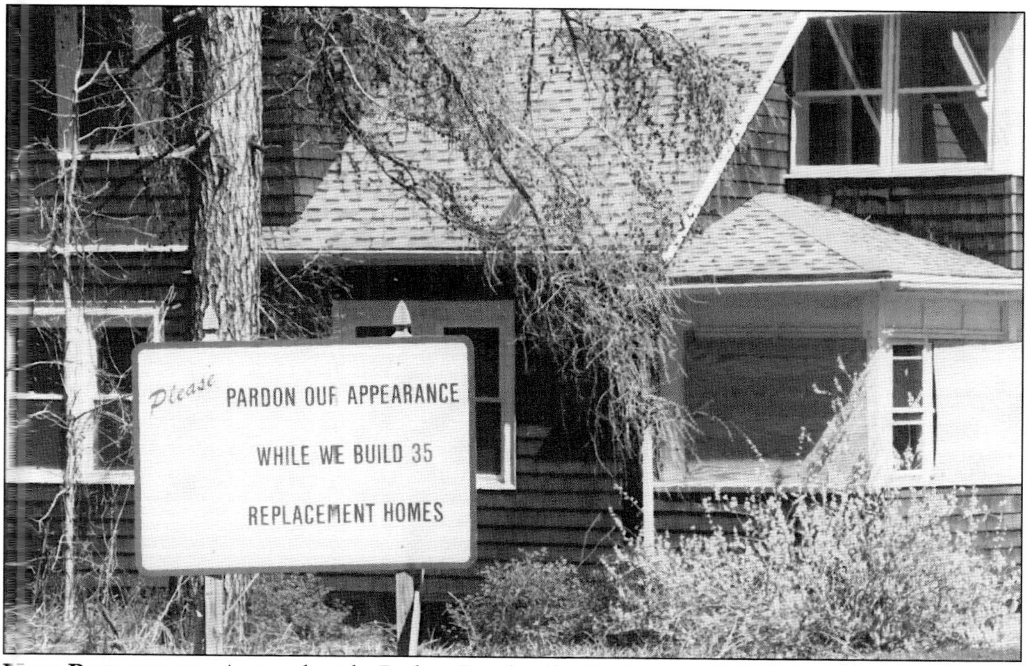

NEW RESIDENCES. A sign beside Parker Road tells the future of the boarded up residences, several of which serve as a backdrop for the sign. In red letters it states: "Please pardon our appearance while we build 35 new replacement homes." Some will be located on Parker Road, others in different sections of the arsenal. (Author's Archives.)

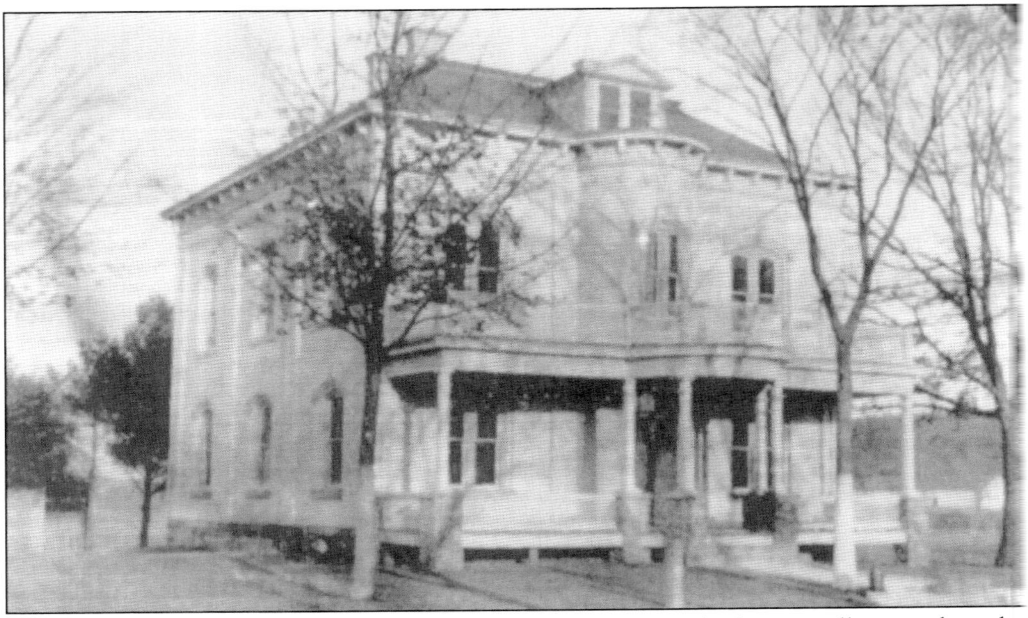

THE CHIEF OF STAFF'S RESIDENCE. This Victorian house, which originally served as the commander's residence, was extensively renovated in 1939. A wooden porch in the front and a chimney were removed, a rear porch replaced, and a wing added. Rotted and broken beams, laths, and plaster were replaced on the interior . (U.S. Army Photograph-Picatinny Arsenal.)

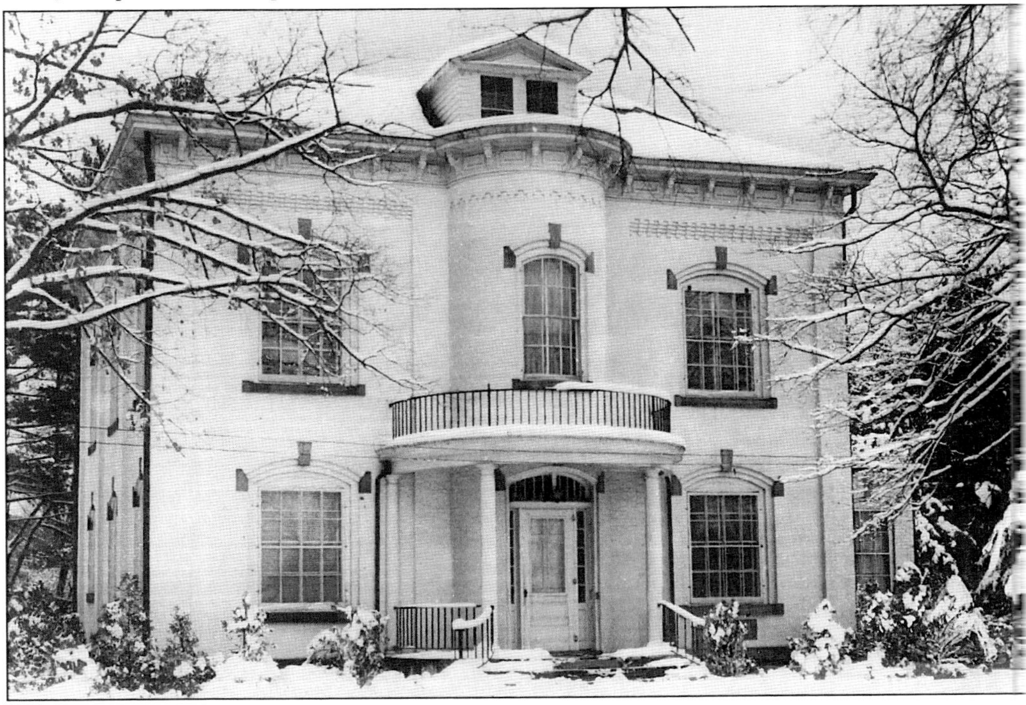

THE ORIGINAL CHIEF OF STAFF'S RESIDENT. This is the residence before the front porch was removed and a wing added on the east facade of the stucco-covered residence. Built in 1884, the two-story residence was valued at $20,000 when built. After reconstruction, the army valued it at $52,458. (U.S. Army Photograph-Picatinny Arsenal.)

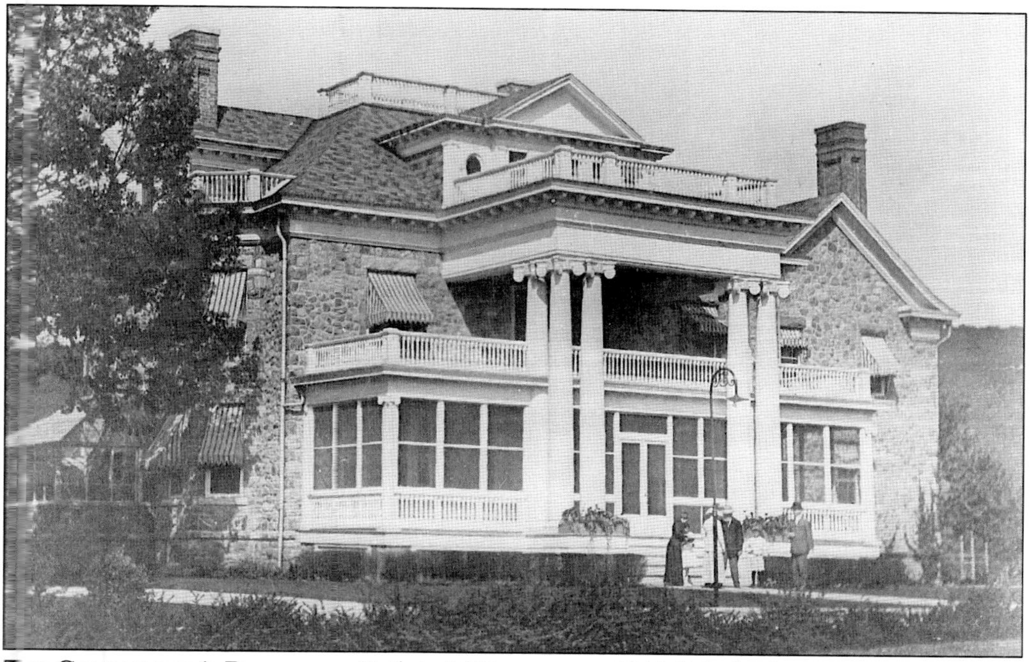

THE COMMANDER'S RESIDENCE. Built in 1909, at a cost of $41,500, this house was first occupied by Maj. O.C. Horney (commander 1907–1915). Facing the parade field, it was constructed of blue puddingstone, a local conglomerate. Its interior was remodeled in 1937 by the WPA. (U.S. Army Photo-Picatinny Arsenal.)

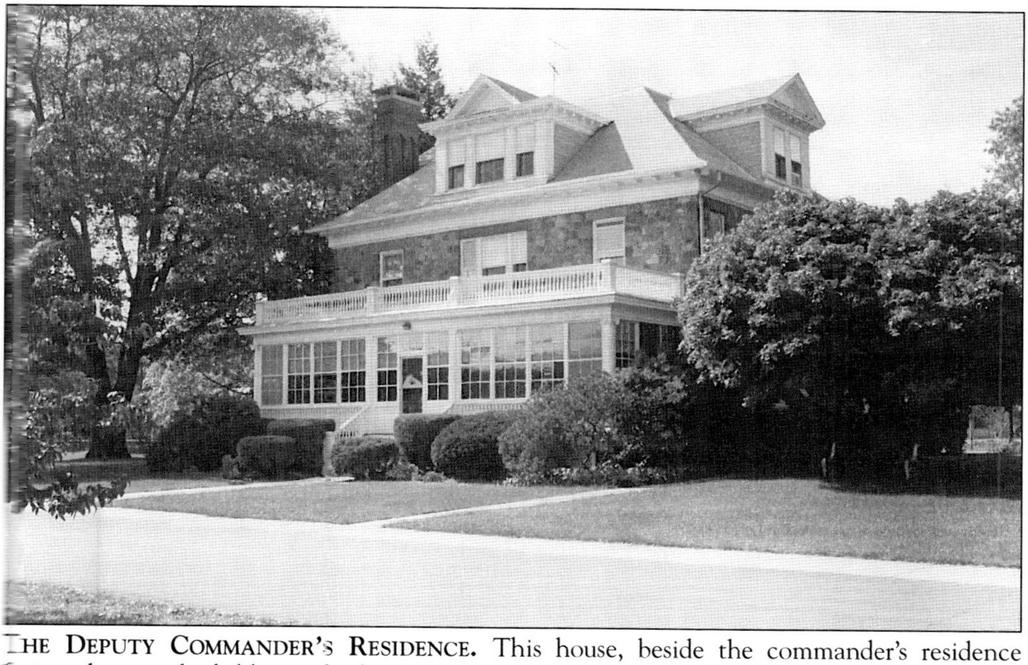

THE DEPUTY COMMANDER'S RESIDENCE. This house, beside the commander's residence facing the parade field, was built in 1909, at a cost of $40,000. Slightly smaller than the commander's residence, it was constructed from blue puddingstone, a local conglomerate. Renovations to the interior in 1938 included the installation of electricity. (U.S. Army Photograph-Picatinny Arsenal.)

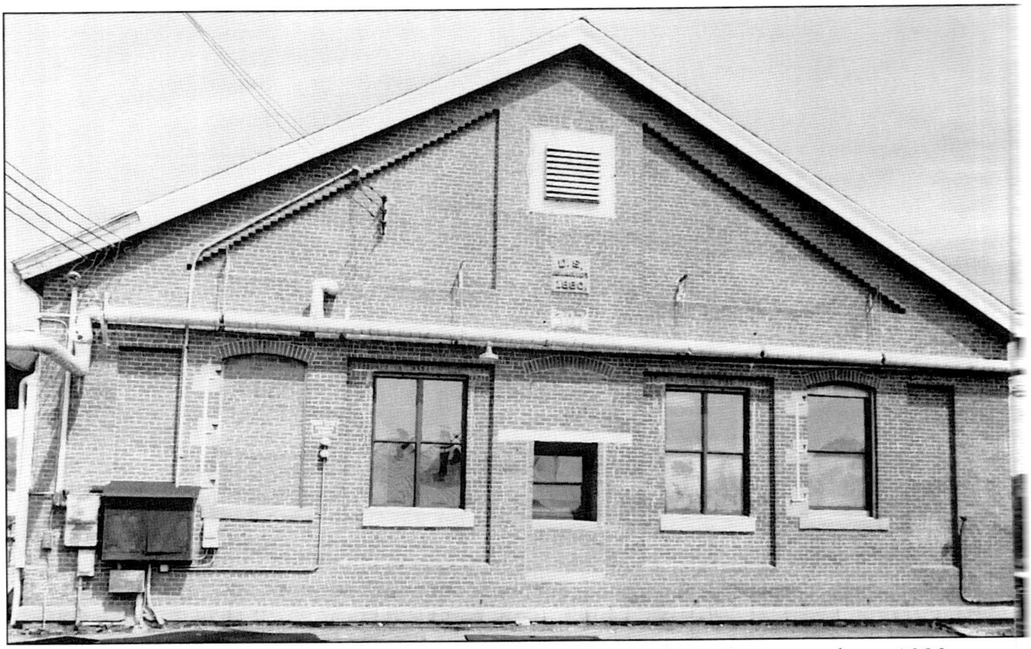

MAGAZINE NUMBER ONE. The first building constructed at the arsenal in 1880 was a 200-by-50-foot brick powder storage magazine with a galvanized iron roof, costing $51,700. It was designed to hold 10,000 barrels of black powder, the only explosive then in use. It was the only magazine with a 6-foot basement to store tools and powders barrels. (U.S. Army Photograph-Picatinny Arsenal.)

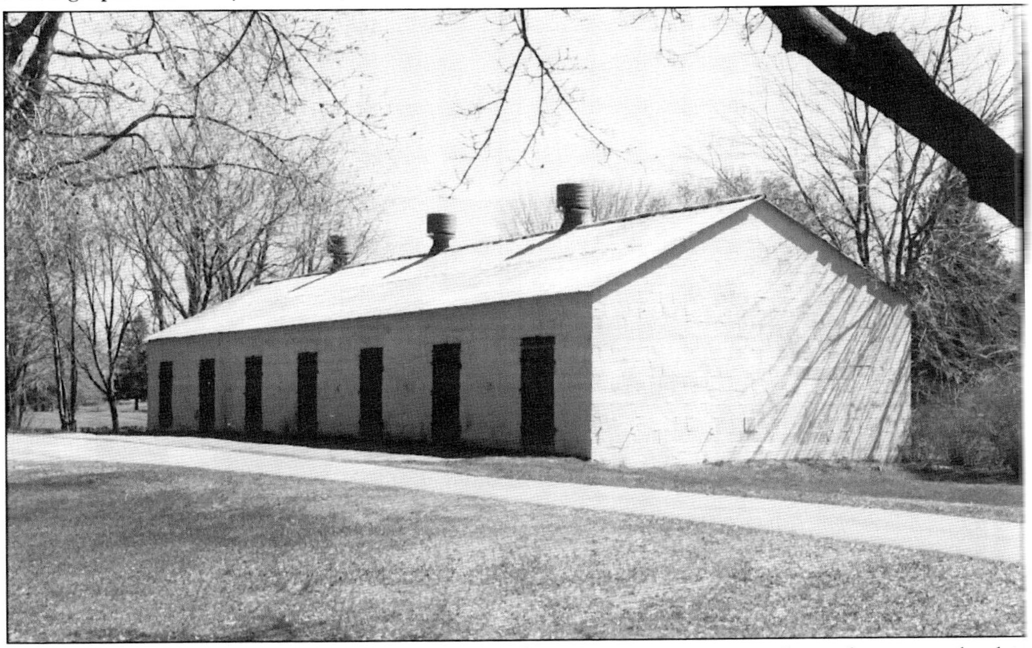

A TYPICAL MAGAZINE. One of the original magazines constructed at the arsenal, this one-story building built in 1918 is typical of early magazines, some of which were constructed of brick, others tile and wood. Sixty-five were rehabilitated by the WPA. (U.S. Army Photograph-Picatinny Arsenal.)

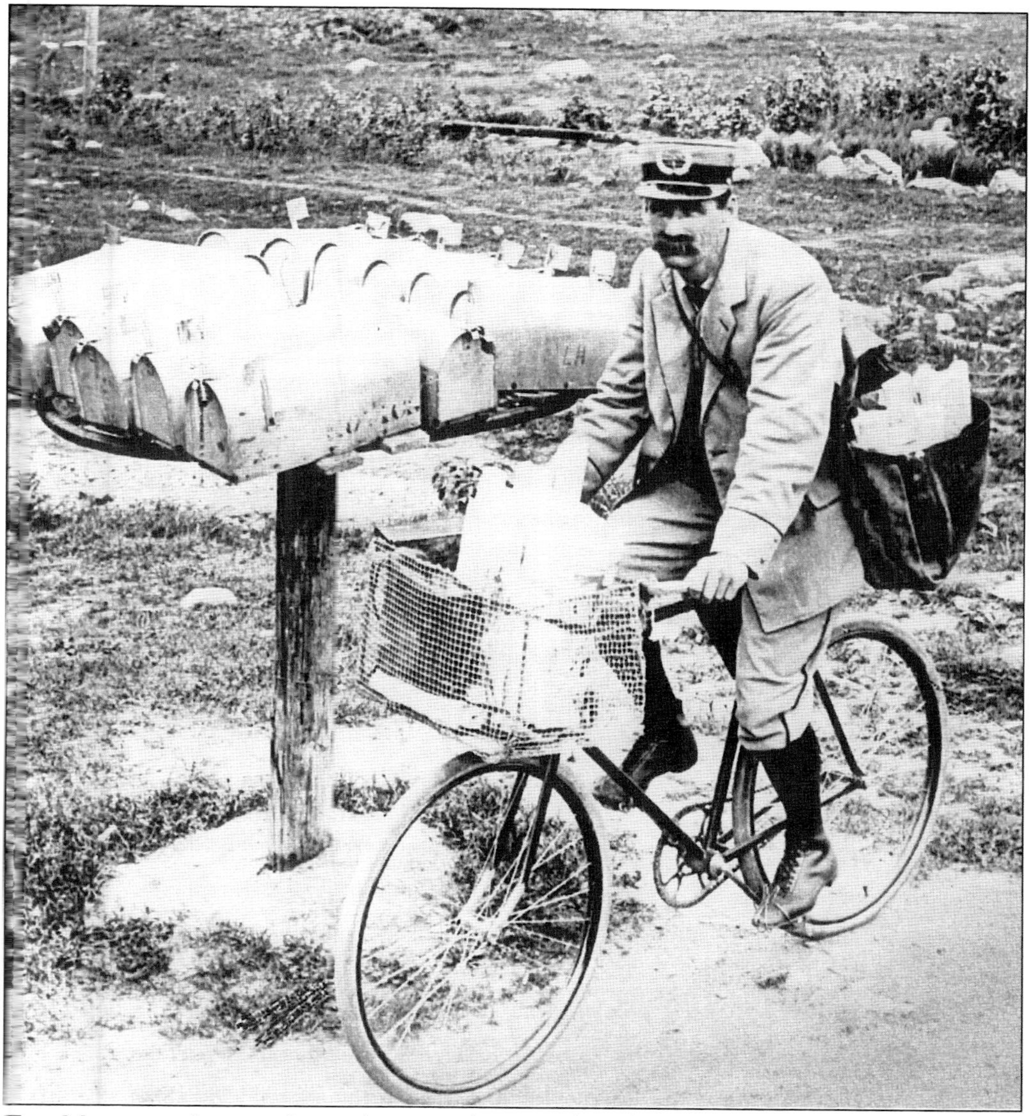

THE MAILMAN. In its infancy, the arsenal received mail from the Wharton Post Office. A mailman peddling a bicycle made what was then known as Route 1, a 25-mile route that served 243 boxes, 14 of which are seen here. Today the arsenal has its own post office. (U.S. Army Photograph-Picatinny Arsenal.)

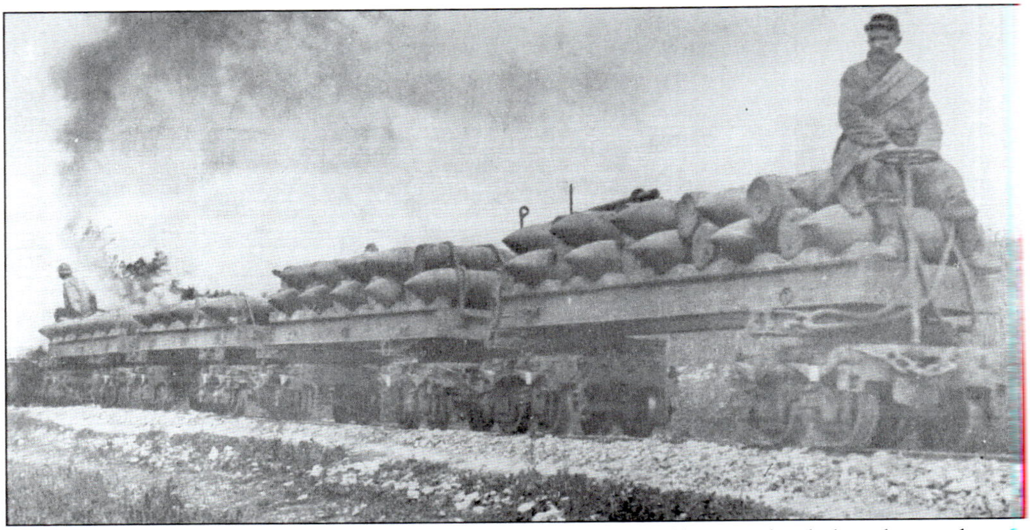

WORLD WAR I SHELLS. During World War I, artillery shells were loaded and stored with smokeless powder, explosives, and metal components in 54 storage magazines, mostly wooden, erected in 1917-18. Large caliber shells are shown being moved between storage buildings on four flatcars. (U.S. Army Photograph-Picatinny Arsenal.)

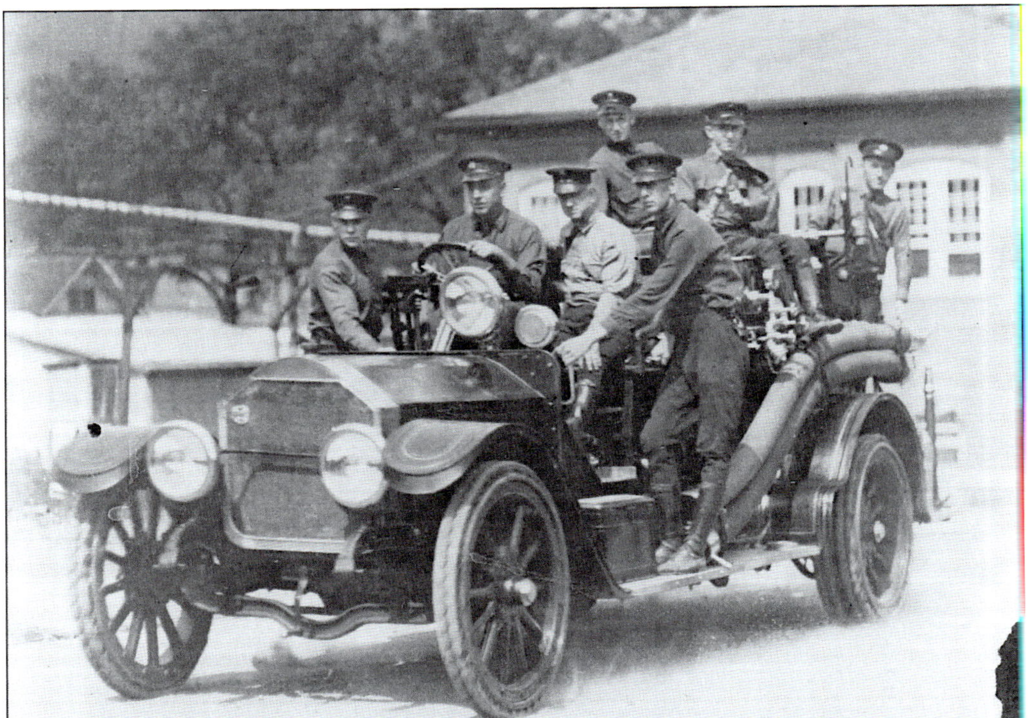

FIREFIGHTING EQUIPMENT. This photograph shows one of the arsenal's earliest fire trucks (1917-18). The firemen mainly extinguished brush fires ignited by sparks from locomotives. In the 1930s, when eight fireless locomotives were in service on the narrow gauge line, 57 miles of railroad, roads, and electrical lines were cleared of brush and trees by the WPA to create fire breaks. (U.S. Army Photograph-Picatinny Arsenal.)

Two

The Big Bang
The Lake Denmark Explosion

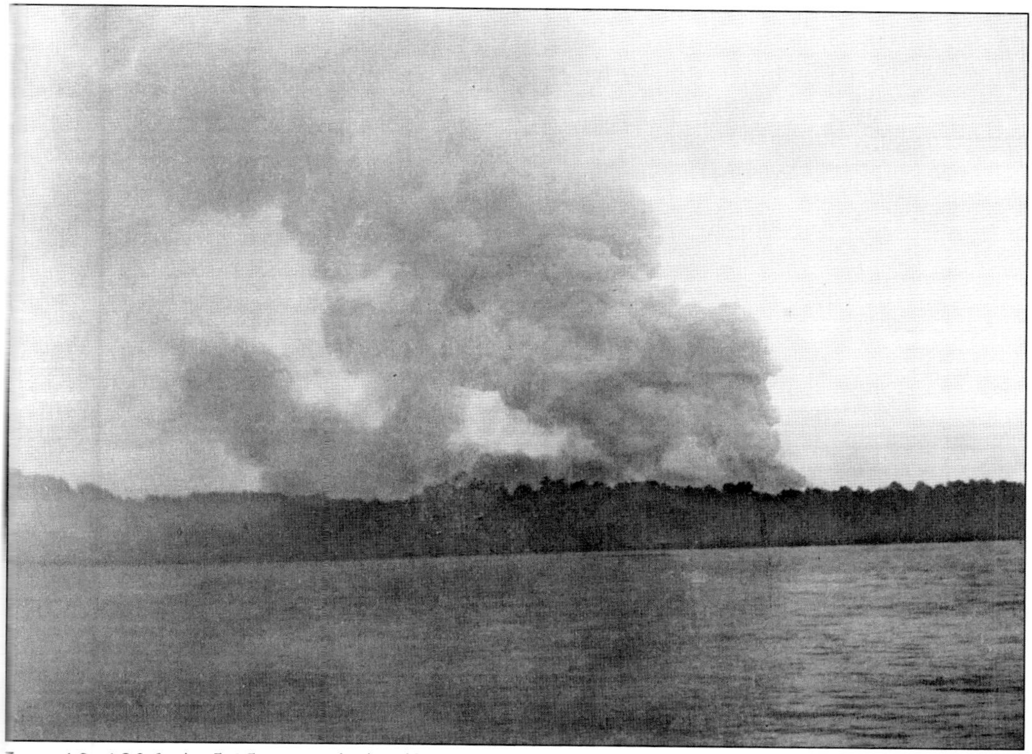

July 10, 1926. At 5:15 p.m., a bolt of lightning struck the U.S. Naval Powder Depot detonating storehouse number 8, where 670,000 pounds of high explosives were stored. This is the towering column of smoke from subsequent explosions and fires as seen from the Sussex County shore of Lake Hopatcong. (U.S. Army Photograph-Picatinny Arsenal.)

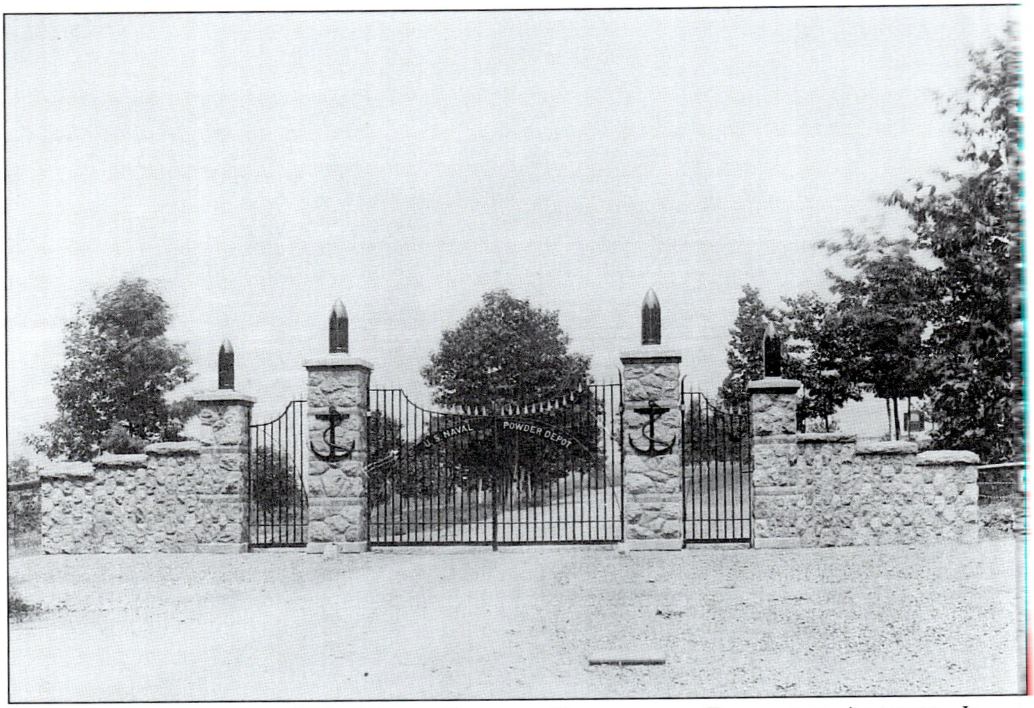

Gates Separating the U.S. Naval Powder Depot from Picatinny Arsenal. It was created in 1891 when the arsenal transferred 315 acres to the navy for construction of magazines 44 of which contained high explosives, smokeless powder, projectiles, and black powder when the explosion occurred. (U.S. Army Photograph-Picatinny Arsenal.)

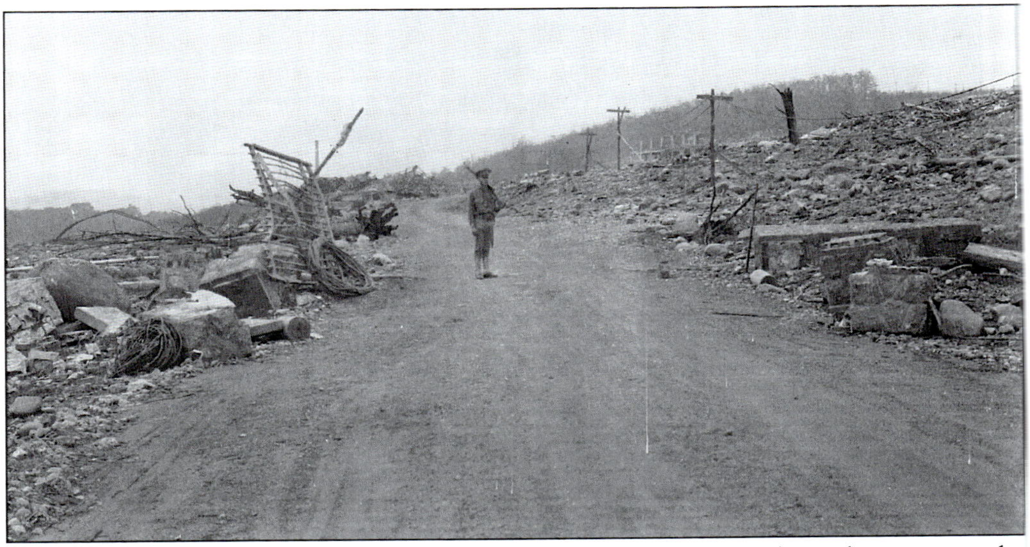

Devastation. After the explosion, a lone Marine stands guard where the gates to the U.S. Naval Powder Depot once stood. Only several telephone poles survived the blast. Everything else was destroyed by the series of explosions, fires, and shock waves that bounced from hill to hill, leveling most buildings at the depot and arsenal. (U.S. Army Photograph-Picatinny Arsenal.)

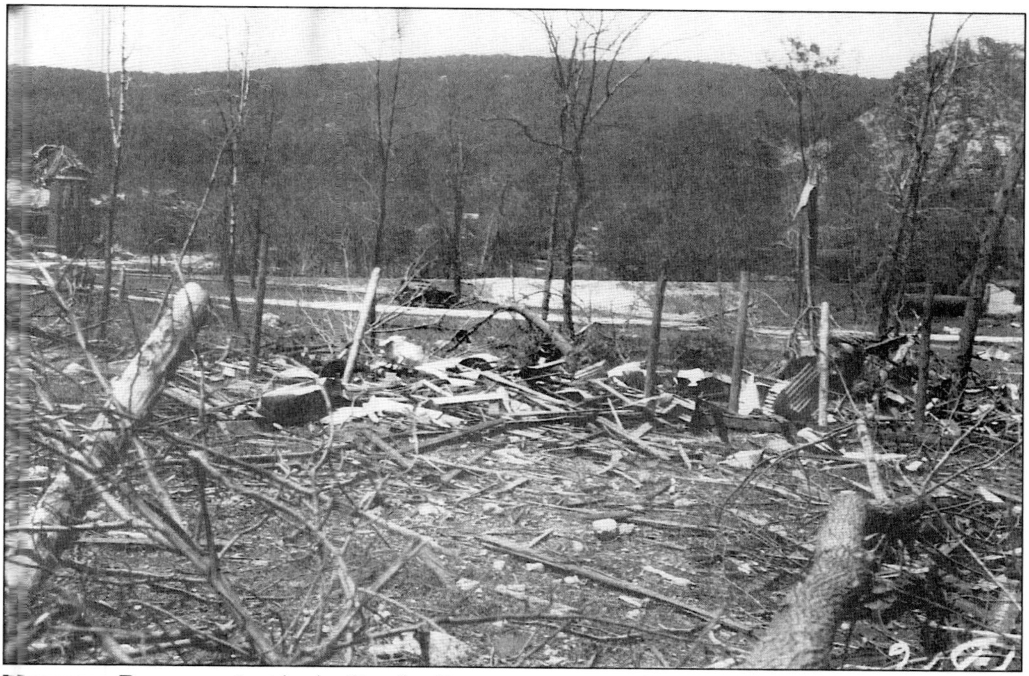

NOTHING REMAINS. Inside the Powder Depot gates, everything was leveled by the explosions; shells landed as far as a mile away in Hibernia, Mount Hope, and Rockaway, where plate glass windows in the business district were blown out. By the time the smoke cleared, 19 persons were dead. Damage was estimated at $1,265,000. (U.S. Army Photograph-Picatinny Arsenal.)

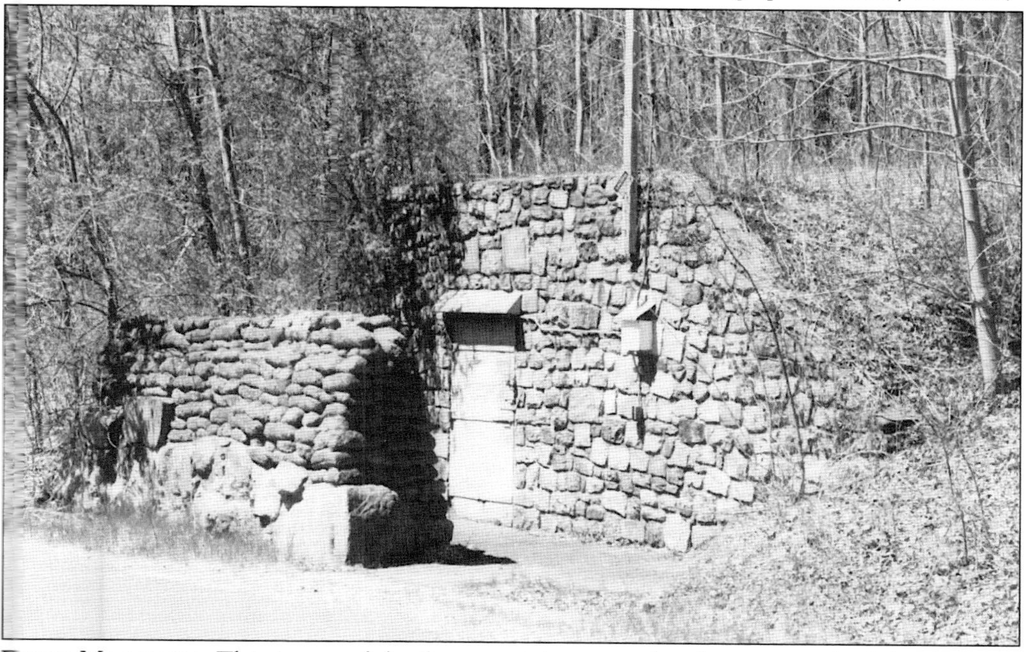

EARLY MAGAZINE. This is one of the few remaining stone-faced magazines built by the navy to store munitions when the Naval Powder Depot was first established. Surrounding the top of Navy Hill, it was one of a series of magazines, many of which were demolished in the 1926 explosion. (Author's Archives)

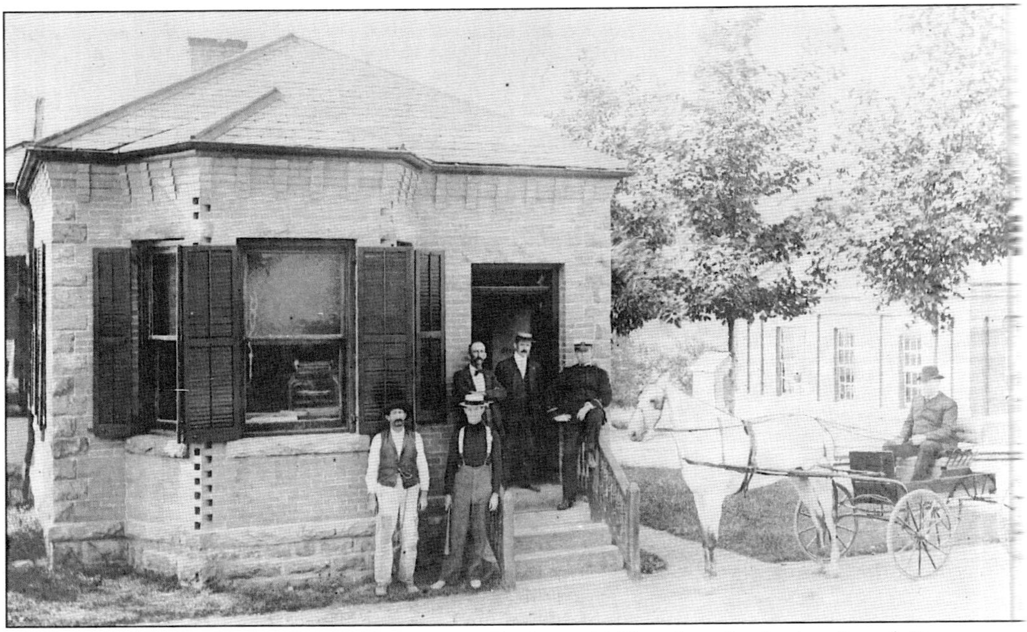

MAIN NAVY ADMINISTRATION BUILDING. Built in 1902, when both the Naval Depot an arsenal had stables for horses and carriages, the main, one-story administration building at th Naval Powder Depot was constructed of brick with granite base and corners. After the explosio it was remodeled and enlarged. (U.S. Army Photograph-Picatinny Arsenal.)

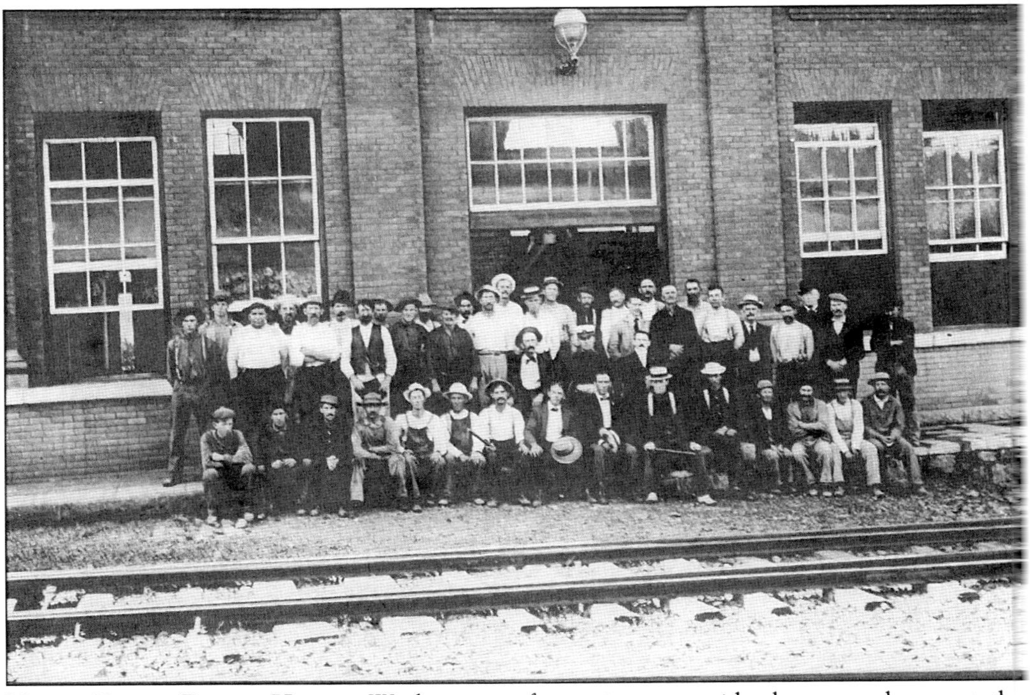

NAVAL DEPOT POWER HOUSE. Workers pose for a picture outside the power house at the Naval Depot. Although protected by the topography from the explosions, its concrete slab roof was crushed and brick walls cracked. A brick stack suffered a 20-foot crack at its top. (U.S. Army Photograph-Picatinny Arsenal.)

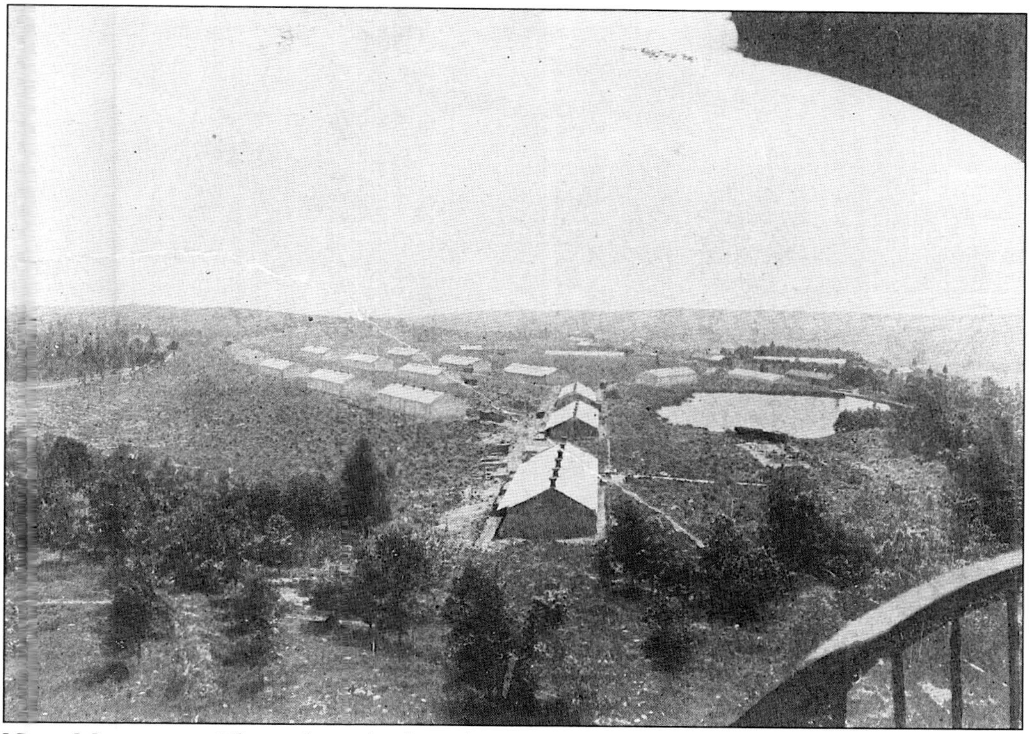

NAVY MAGAZINES. This is the only photo known to exist of the rows of magazines constructed by 1920 at the Naval Depot. They stored everything from small arms ammunition to 16 inch shells for coast defense guns, battleships, mines, and black and smokeless powder. Many of the 44 buildings that housed munitions were described as temporary. (U.S. Army Photograph-Picatinny Arsenal.)

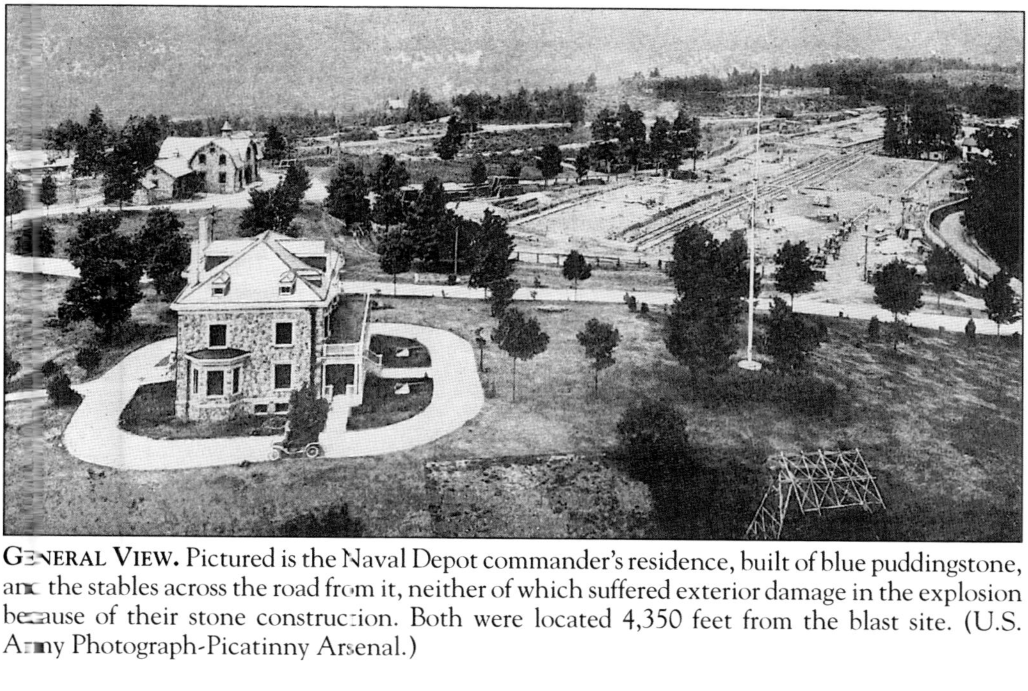

GENERAL VIEW. Pictured is the Naval Depot commander's residence, built of blue puddingstone, and the stables across the road from it, neither of which suffered exterior damage in the explosion because of their stone construction. Both were located 4,350 feet from the blast site. (U.S. Army Photograph-Picatinny Arsenal.)

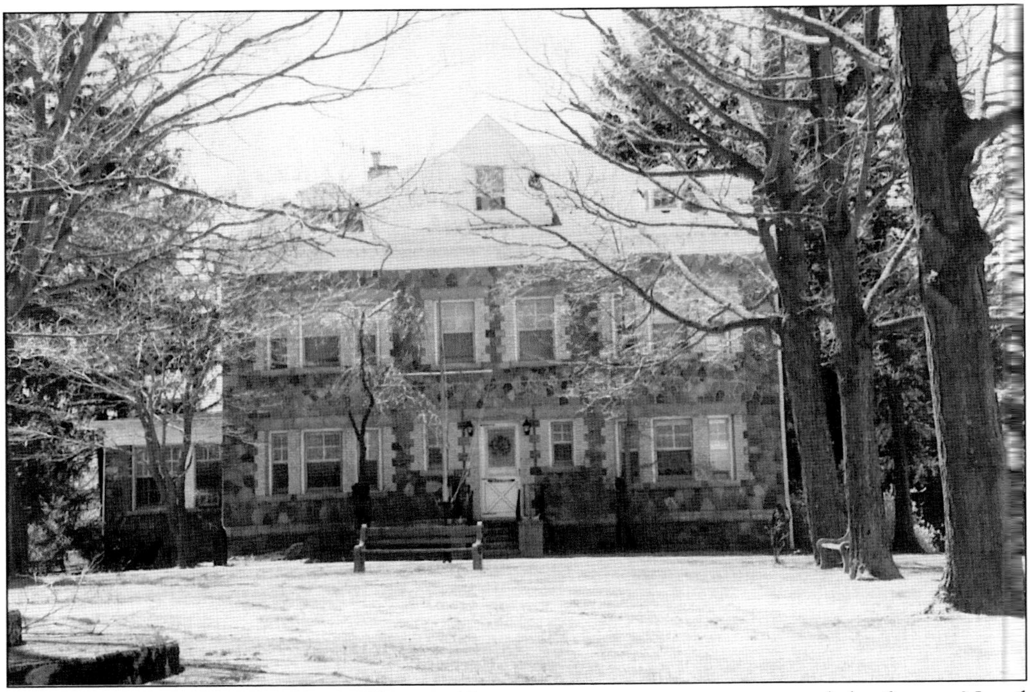

The Commander's Residence Today. Towering evergreen trees surround the former Naval Depot Commander's residence today. It overlooks the stables', now the arsenal firehouse; rear entrance to the arsenal; and buildings stretching along Lake Denmark Road. It is used as officers' quarters. (Author's Archives.)

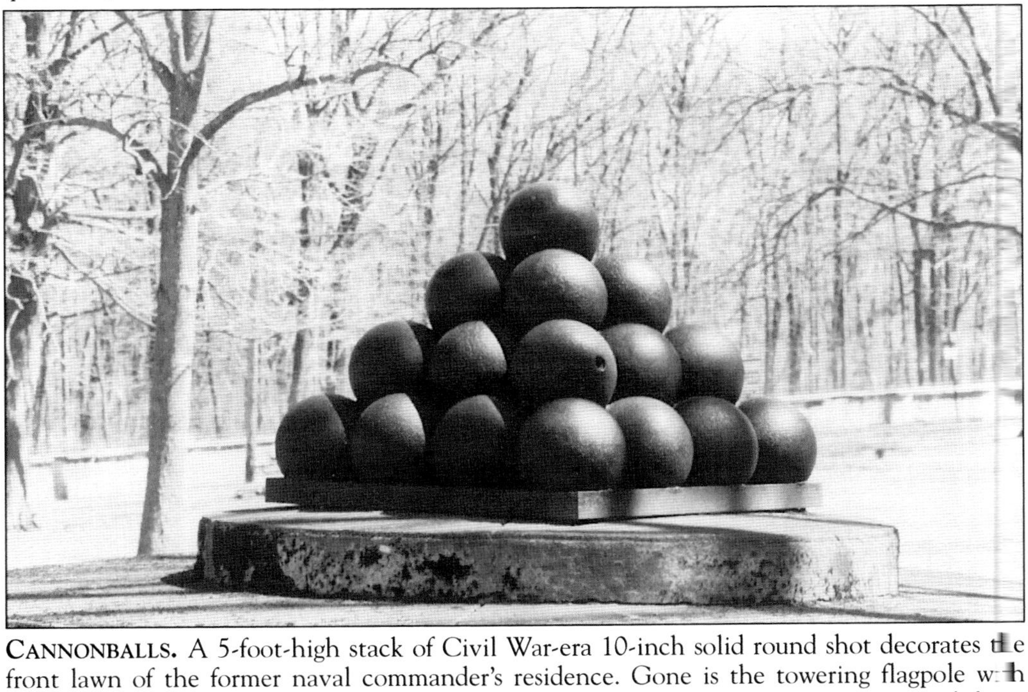

Cannonballs. A 5-foot-high stack of Civil War-era 10-inch solid round shot decorates the front lawn of the former naval commander's residence. Gone is the towering flagpole with crossarms, which once stood beside the residence, and two cannon. Large caliber coast defense gun shells remain at the front corners of the residence. (Author's Archives.)

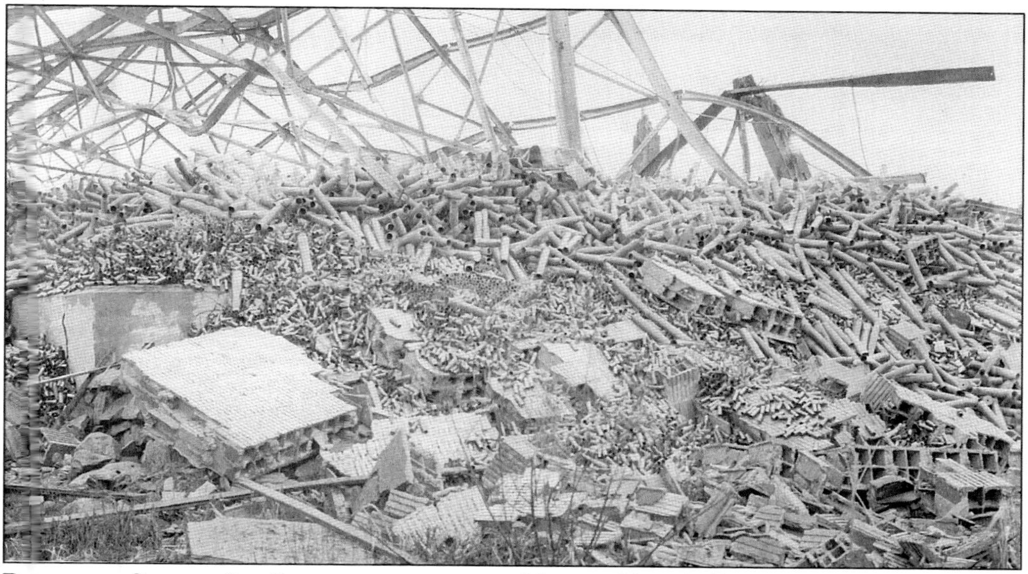

PLENTY OF SHELLS. Small arms ammunition and large caliber shells are strewn throughout the wreckage of this tile magazine, whose side walls collapsed beneath the twisted steel girders that once supported the roof. It was one of 30 magazines in the north area of the depot, 4,000 to 5,600 feet from the blast site. (U.S. Army Photograph-Picatinny Arsenal.)

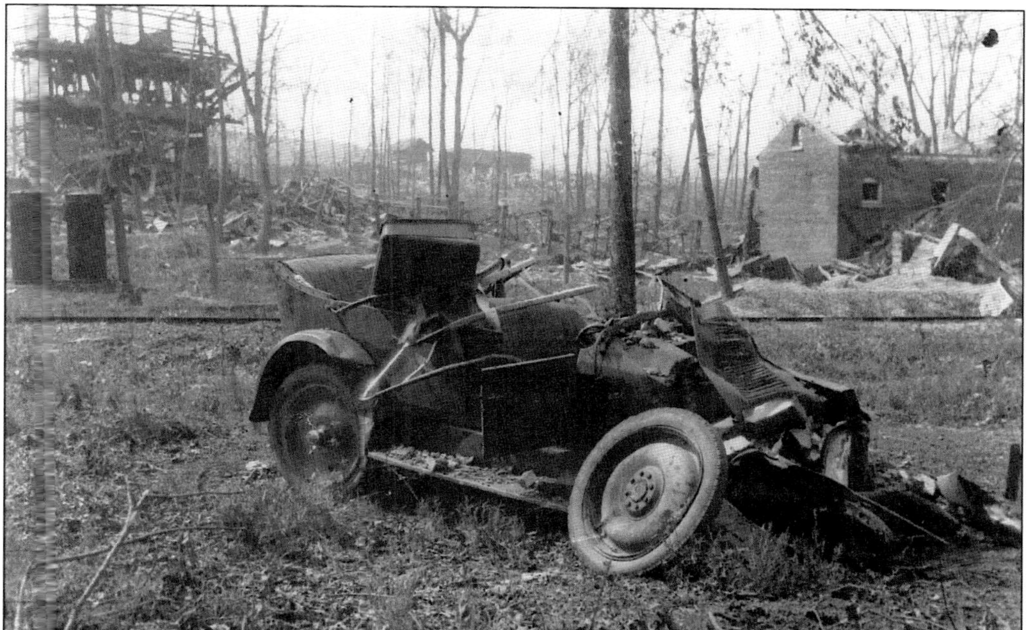

ONCE A CAR. Crushed by the force of the explosion and shock waves that broadened out as they criss-crossed the valley, exceeding 100 miles-an-hour in some areas, this car was thrown against a tree. An army report states the waves exerted a pressure of 50 to 60 pounds per square foot. (U.S. Army Photograph-Picatinny Arsenal.)

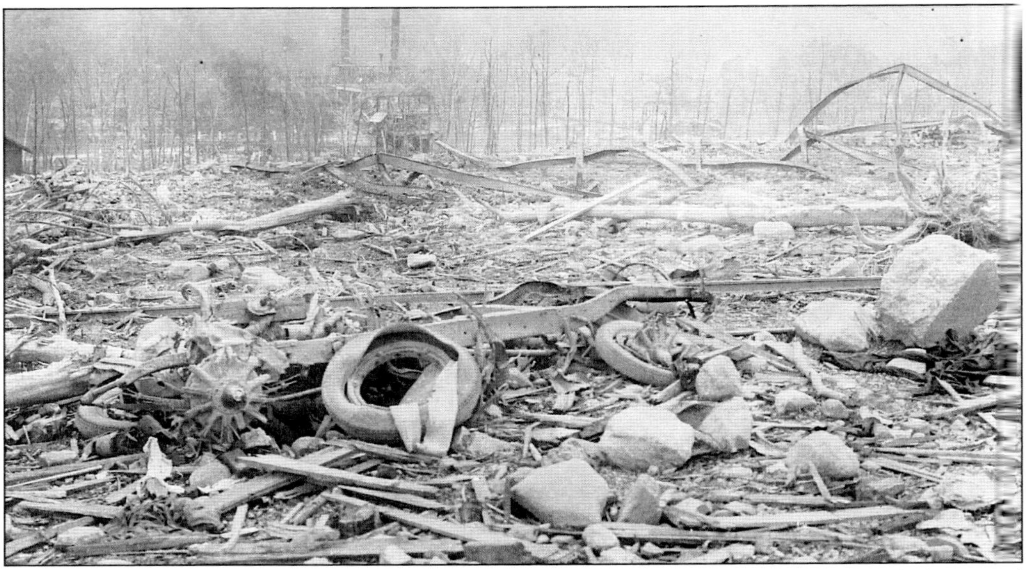

NOT A COLLISION. The shock waves following the explosions uprooted trees, left twisted remains of girders that once supported the roofs of buildings, and scattered the frame, tires, and engine of this vehicle over a wide area. Trees in the background were snapped in half. (U.S. Army Photograph-Picatinny Arsenal.)

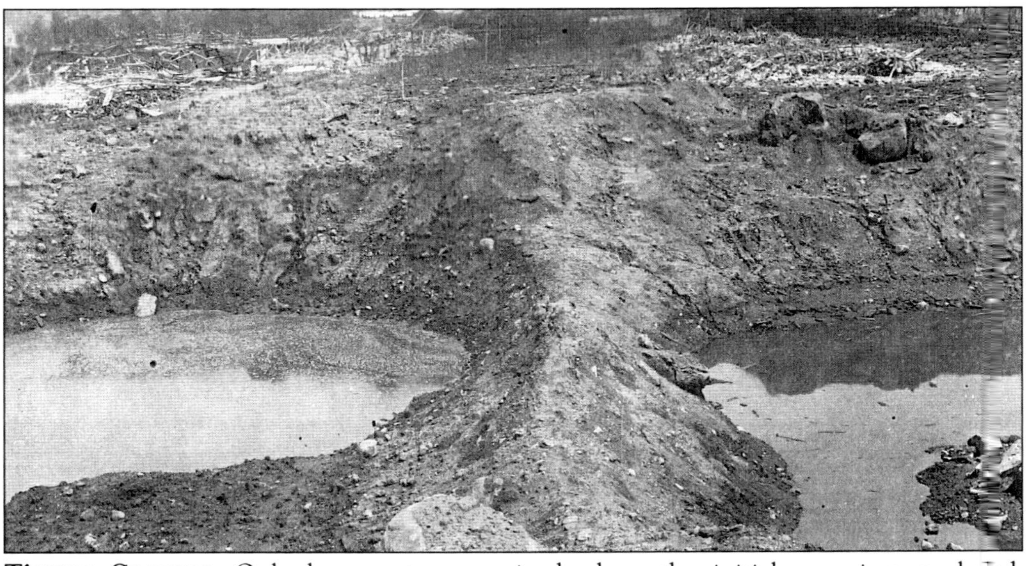

T'WEEN CRATERS. Only large craters remained where the initial magazines exploded. Smaller craters were everywhere. One crater was from an exploded 5-inch shell on the arsenal parade ground, 1 mile from the explosion site; another was from an exploded shell beside the arsenal road to Mt. Hope, 3,000 feet from the explosion site. (U.S. Army Photograph-Picatinny Arsenal.)

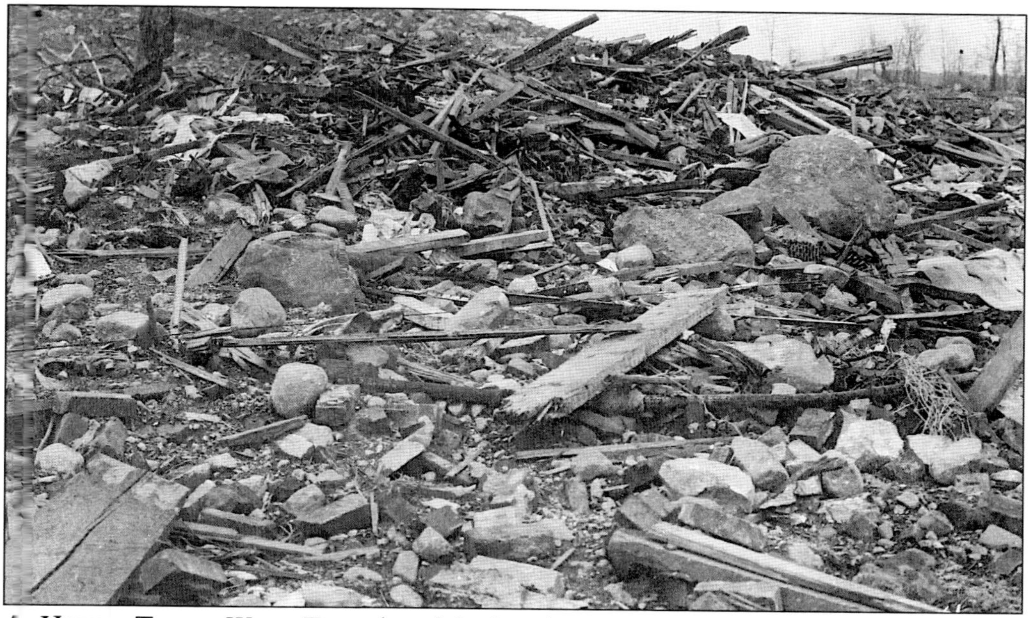

A HOUSE THERE WAS. Twisted and broken beams inter-mixed with smashed window frames and boulders were all that remained of a house in the explosion area where employee homes formed semi-circles around the exploding magazines. All were demolished, and clothing, furniture, and household articles scattered over a wide area. (U.S. Army Photograph-Picatinny Arsenal.)

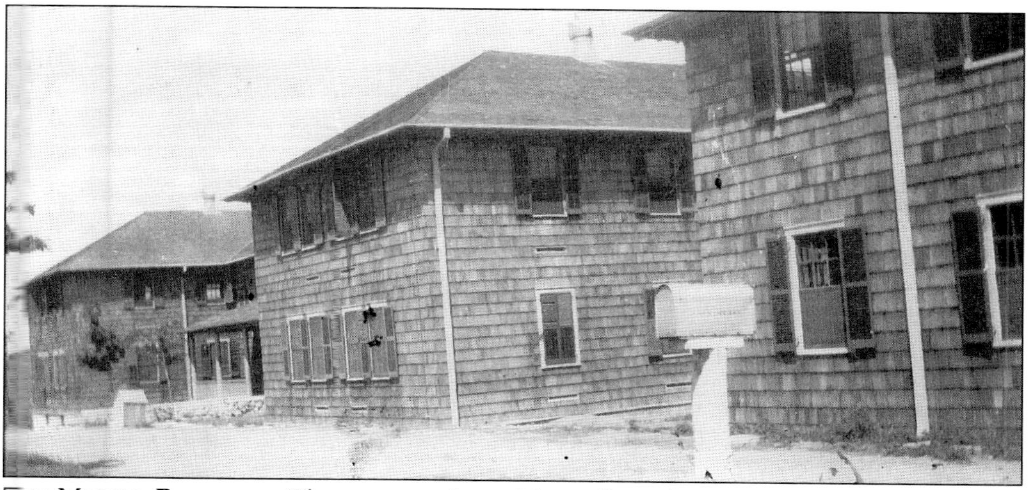

THE MARINE BARRACKS. The two-story frame Marine Barracks, 3,175 feet from the explosion site, was wrecked by the blast. It eventually caught fire and burned to the ground 17 hours after the first explosion. (U.S. Army Photograph-Picatinny Arsenal.)

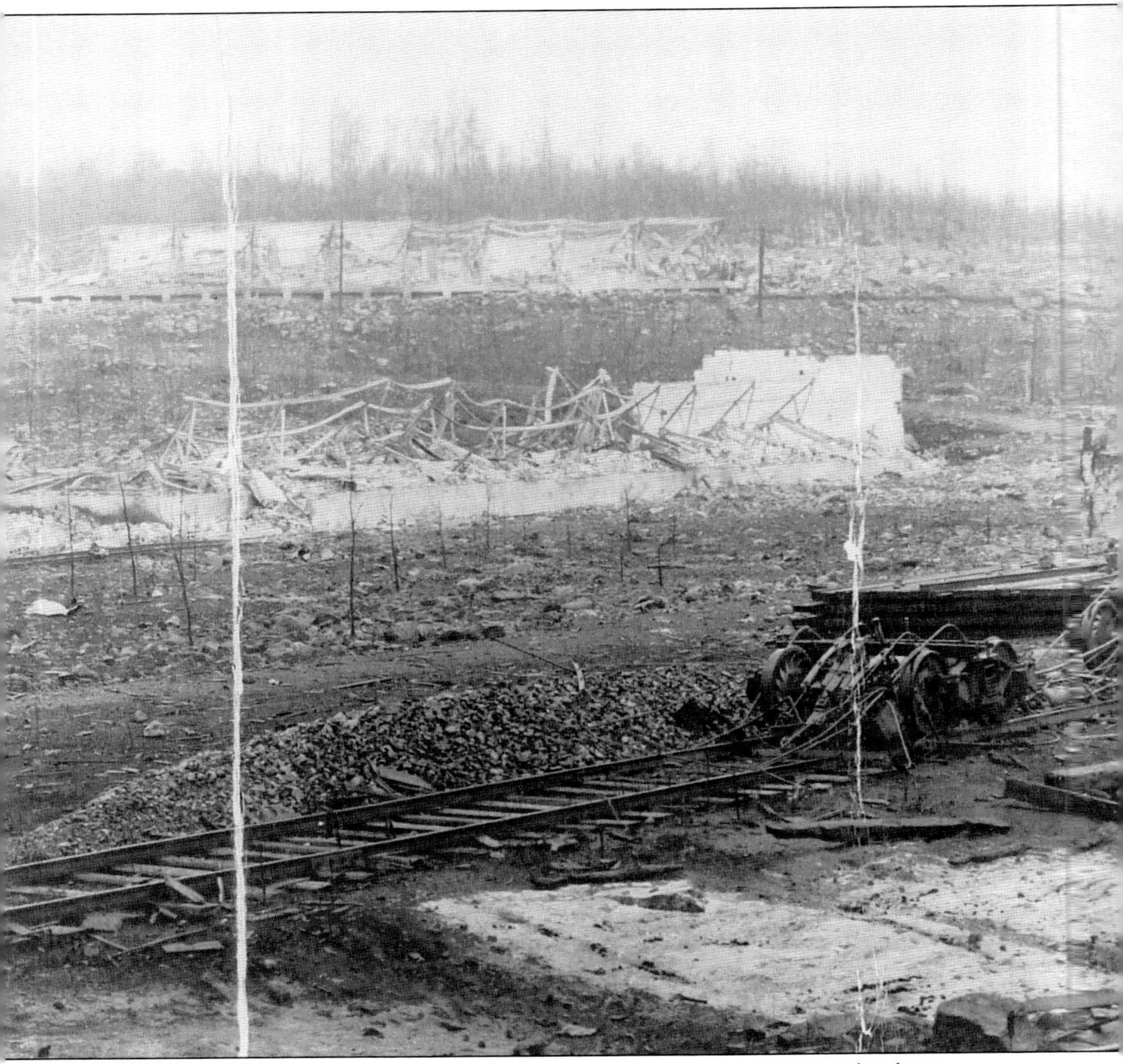

A PANORAMA OF DESTRUCTION. The first explosion of 670,000 pounds of munitions ripped buildings apart, exposing rows of large caliber shells to flaming debris from the second explosion. Pieces of trusses and girders, weighing from a few to several hundred pounds, which supported the roofs of the temporary store houses, were thrown 3,000 to 4,000 feet, many falling on the arsenal reservation. (U.S. Army Photograph-Picatinny Arsenal.)

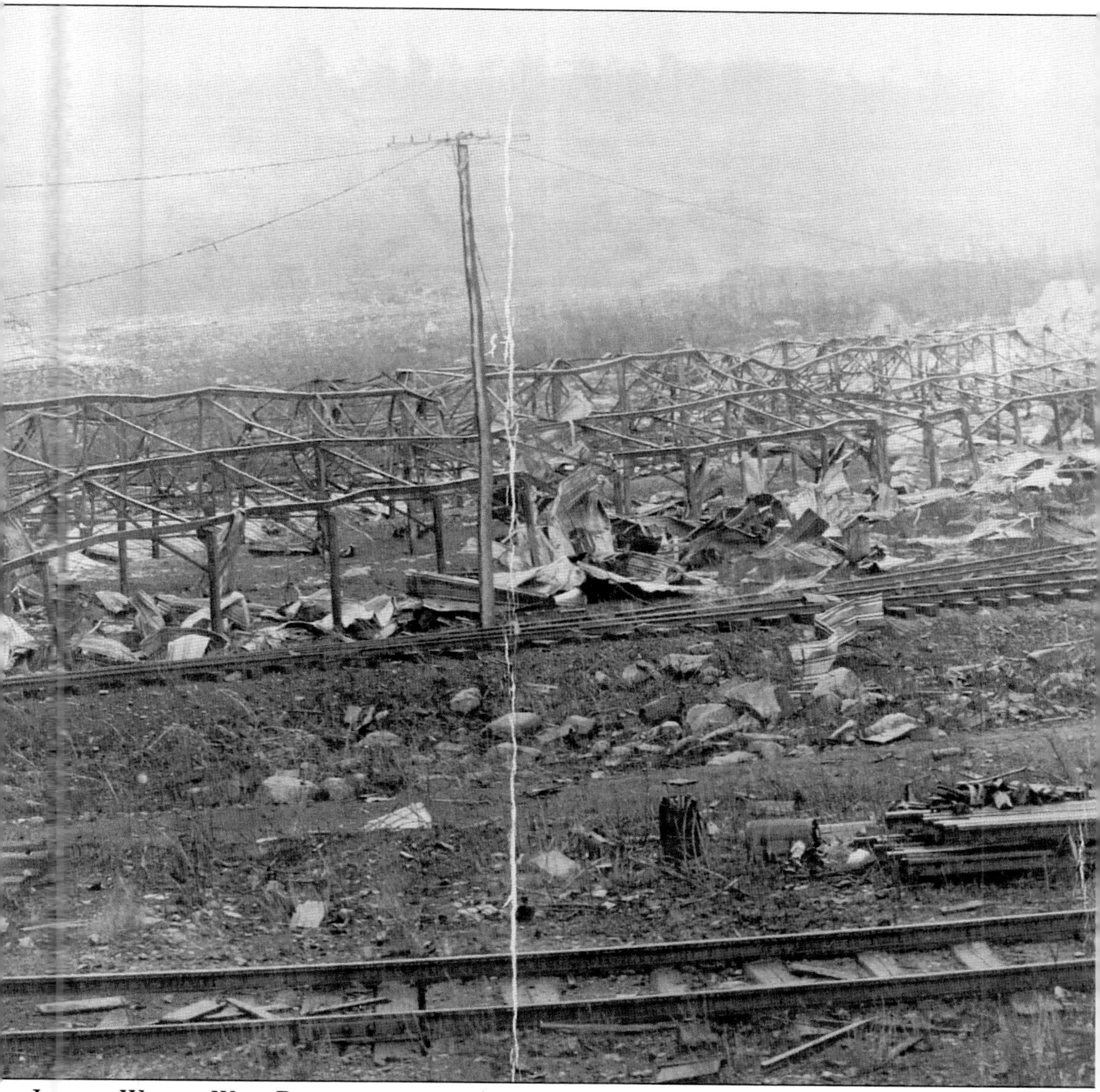

LIKE A WORLD WAR BATTLEGROUND. The remains of a railroad car sit on a track that escaped upheaval in the explosion and following shock waves. The force of the explosions, the biggest of which occurred in storehouse number 9, where 1,600,000 pounds of TNT were stored, scrambled small arms, aircraft bombs, mines, and artillery shells in massive piles beneath twisted girders, demolishing everything within 3,000 feet of the blast site. (U.S. Army Photograph-Picatinny Arsenal.)

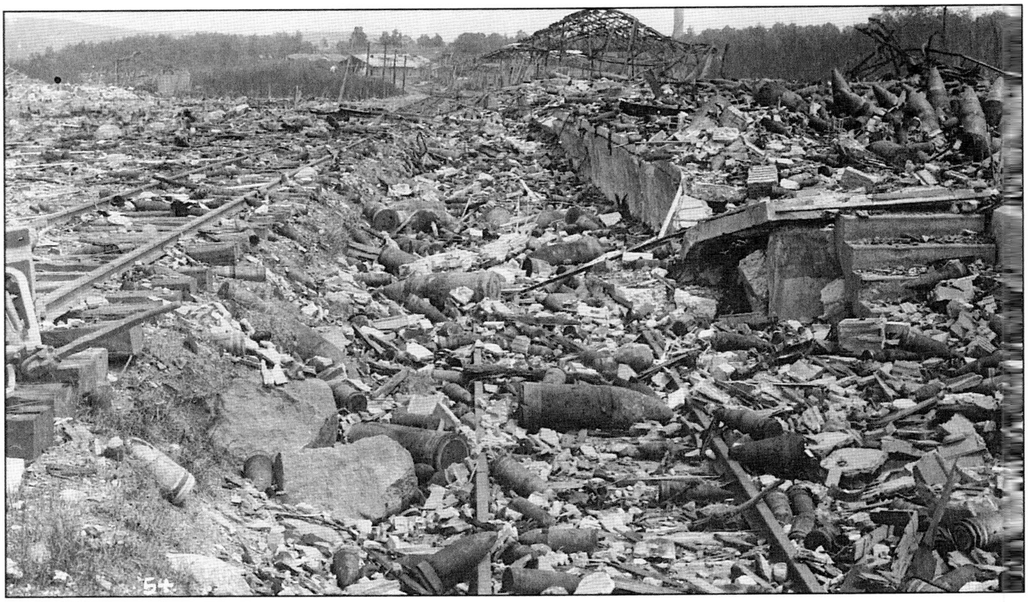

On Dangerous Ground. Utter devastation surrounded the initial magazine that exploded at the Naval Depot. Pictured are twisted railroad tracks covered with unexploded shells of varying caliber. On the right, steps lead to what was a storage shed, its floor heaped with piles of large caliber shells. (U.S. Army Photograph-Picatinny Arsenal.)

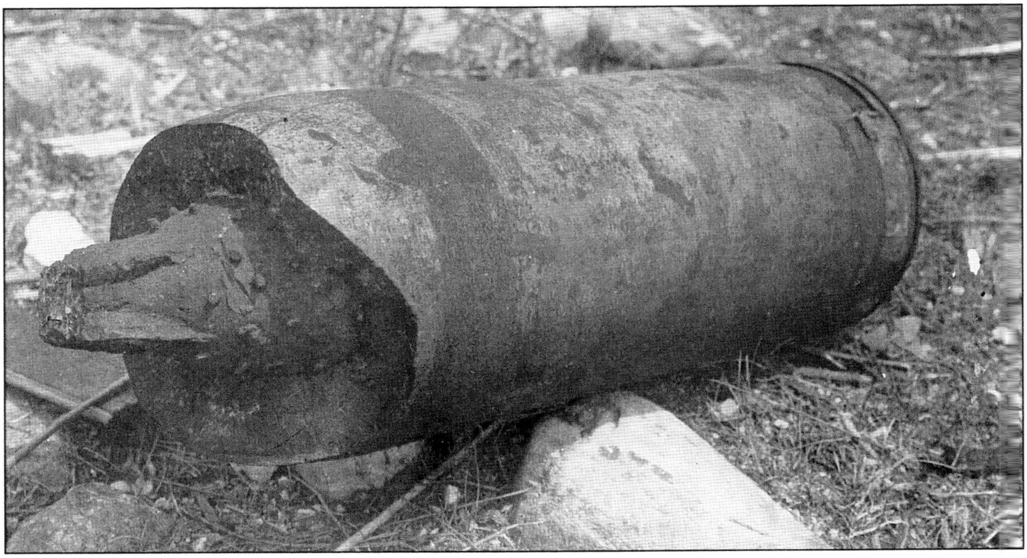

Large Shell. A large shell with its nose cone ripped off by the force of the explosion lies amid scattered rubble. Everything within a 3,000-foot radius of the first two magazines that exploded was wrecked, burned, or destroyed, while many buildings beyond 4,000 feet suffered only minor damage. (U.S. Army Photograph-Picatinny Arsenal.)

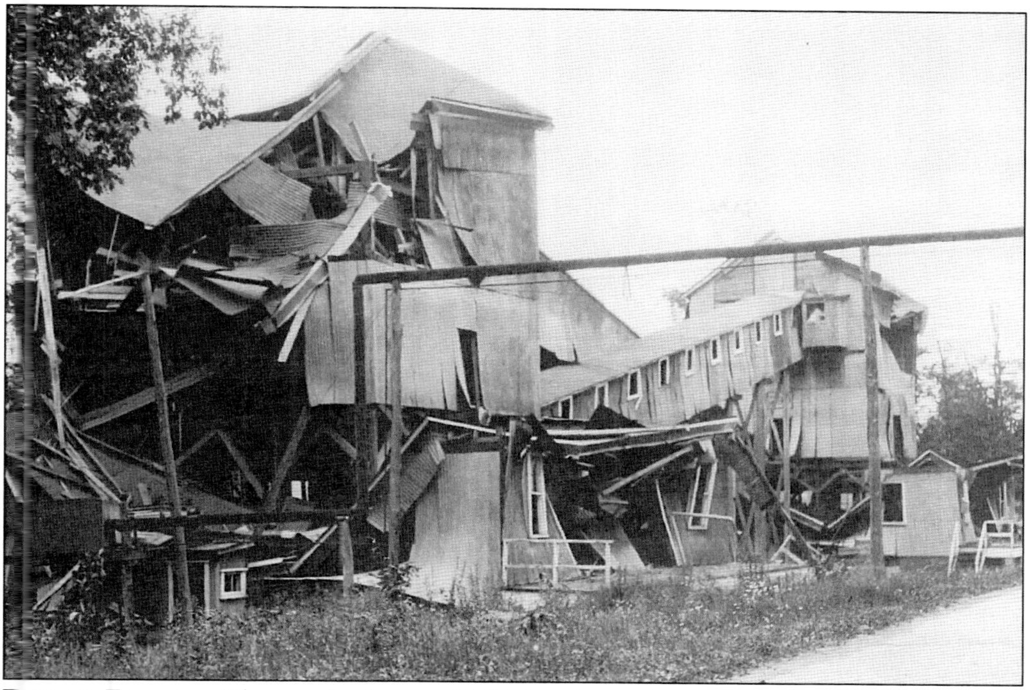

POWDER FACTORY. The powder factory, 2,750 feet distant from the explosion site, was badly wrecked but not entirely demolished. Like nearby powder, drying, and blending houses, part of its roof collapsed. Its frame was twisted and one end of the three-story building, dating to 1907, ripped off. (U.S. Army Photograph-Picatinny Arsenal.)

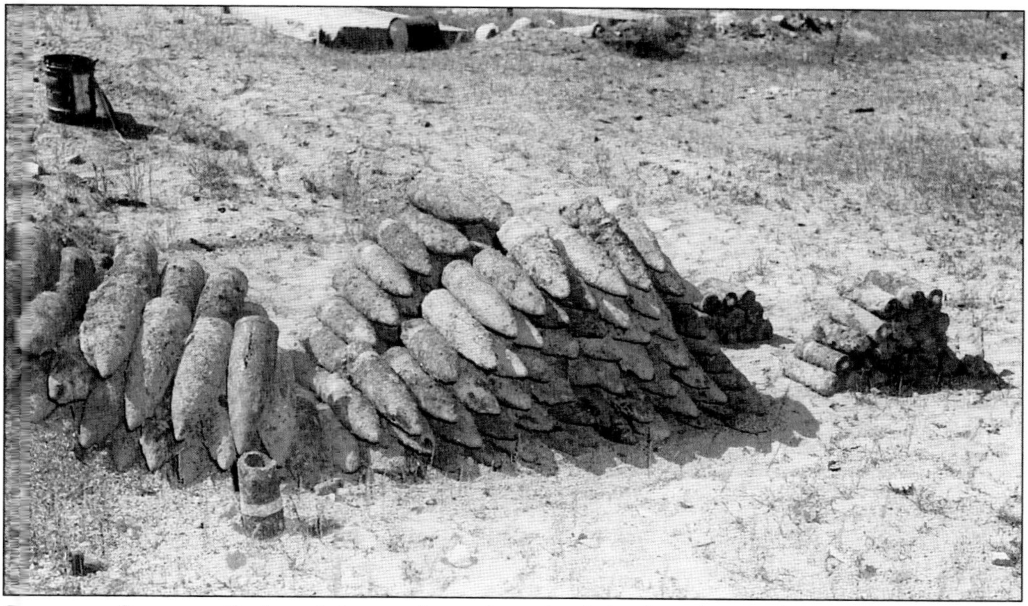

STACKED SHELLS. Gathered from where they landed after the explosion, artillery shells were stacked in the open for lack of covered storage space. Immediately after the explosion, Congress authorized the arsenal's rebuilding. In the mid-1930s, the WPA spent $2,953,920 on rehabilitation, during World War II another $8 million was spent, and in 1945, $7 million. (U.S. Army Photograph-Picatinny Arsenal.)

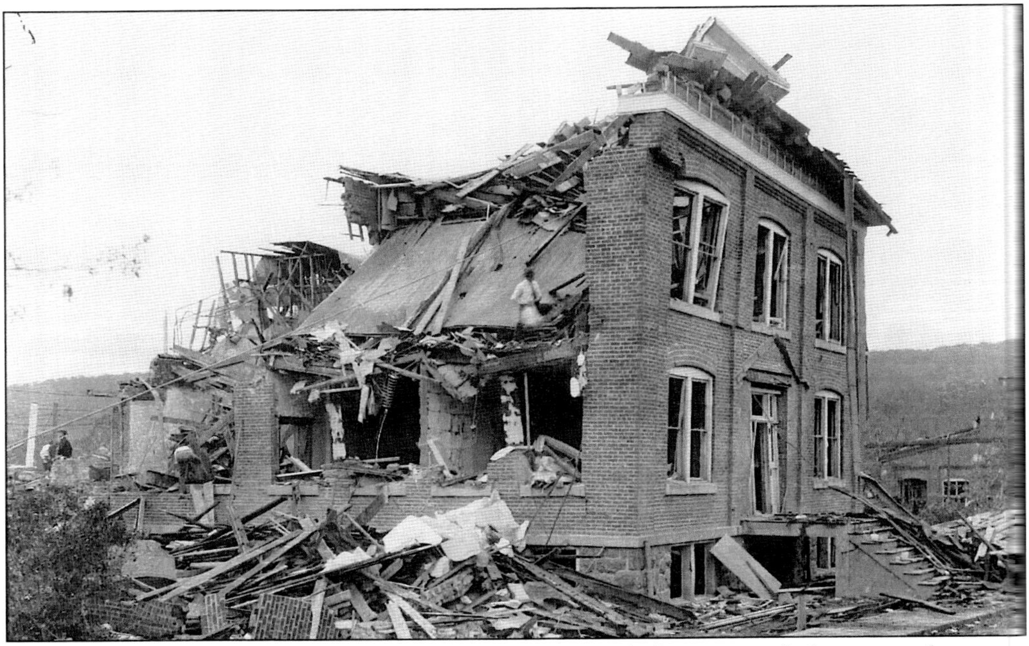

The Administration Building. This is a view of the ruins of the original arsenal administration building, its rear portion demolished and roof caved in by the force of the explosion and resulting shock waves. It was replaced in 1929 with a larger two-story brick command center at a cost of $150,000. This building, with a subsequent addition, served as the core of administration in World War II. (U.S. Army Photograph-Picatinny Arsenal.)

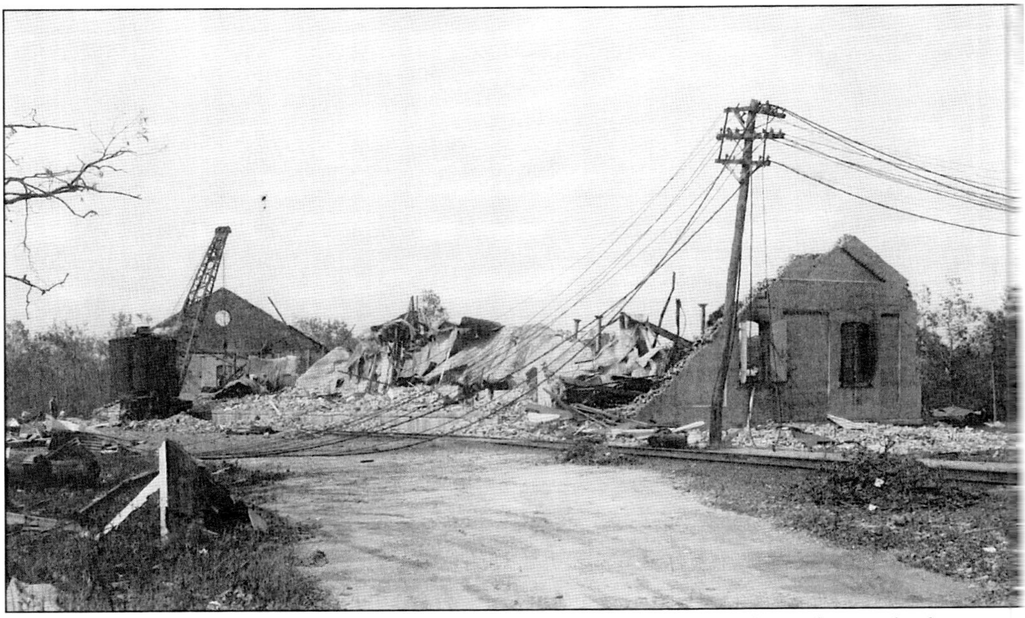

The Clean-up Begins. A crane begins the job of removing debris from the crushed remains of a building at the arsenal. The force of the repeated explosions and shock waves moved the twisted remains of buildings onto railroad tracks, demolishing railroad cars and engines. Brick walls were glazed by the intense heat that melted many of the bricks. (U.S. Army Photograph-Picatinny Arsenal.)

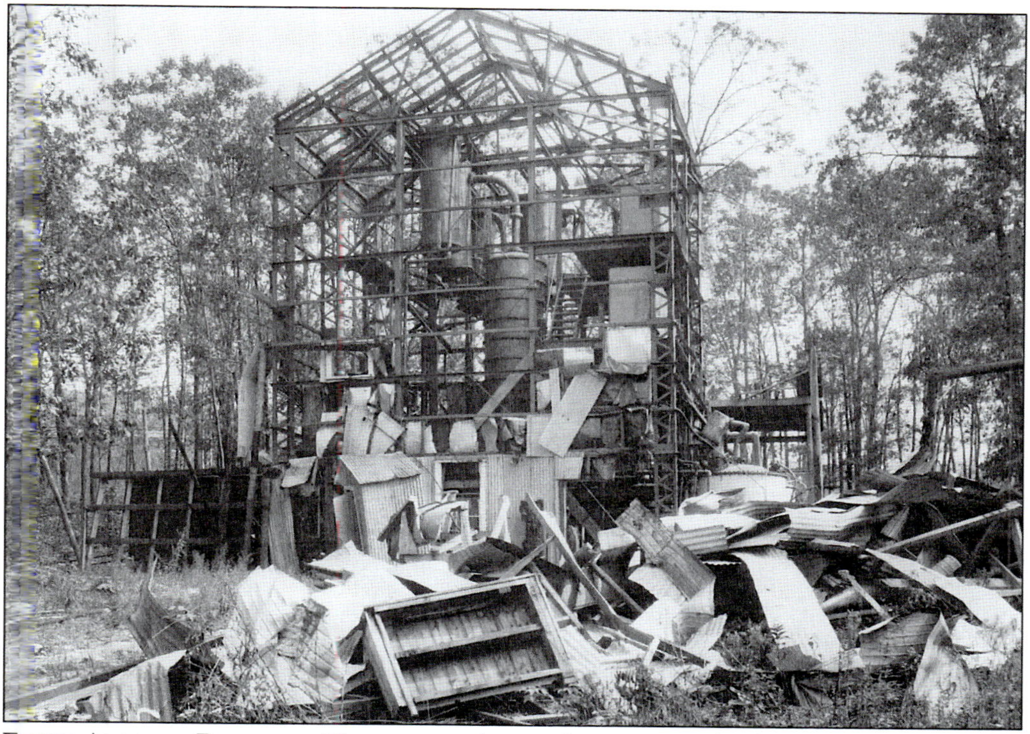

ETHER ALCOHOL BUILDING. The corrugated iron sides were completely ripped off and the steel frame twisted and bent by the force of the explosions and resulting shock waves. This relatively high building, 1,650 feet from the blast site, was built in 1907 at a cost of $3,542. Machinery inside suffered little damage. (U.S. Army Photograph-Picatinny Arsenal.)

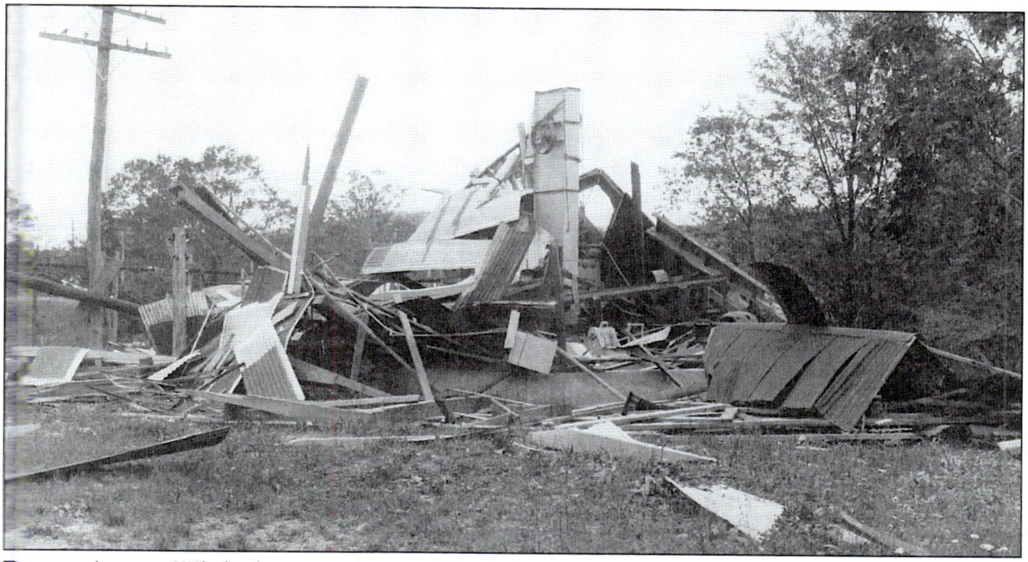

RIPPED APART. While the central core of this Lake Denmark building and an adjacent telephone pole survived the explosion, the outer walls were ripped apart by the force of the explosions and resultant shock waves. Sheets of corrugated metal, wooded structural beams, and the partially intact roof were scattered over a wide area. (U.S. Army Photograph-Picatinny Arsenal.)

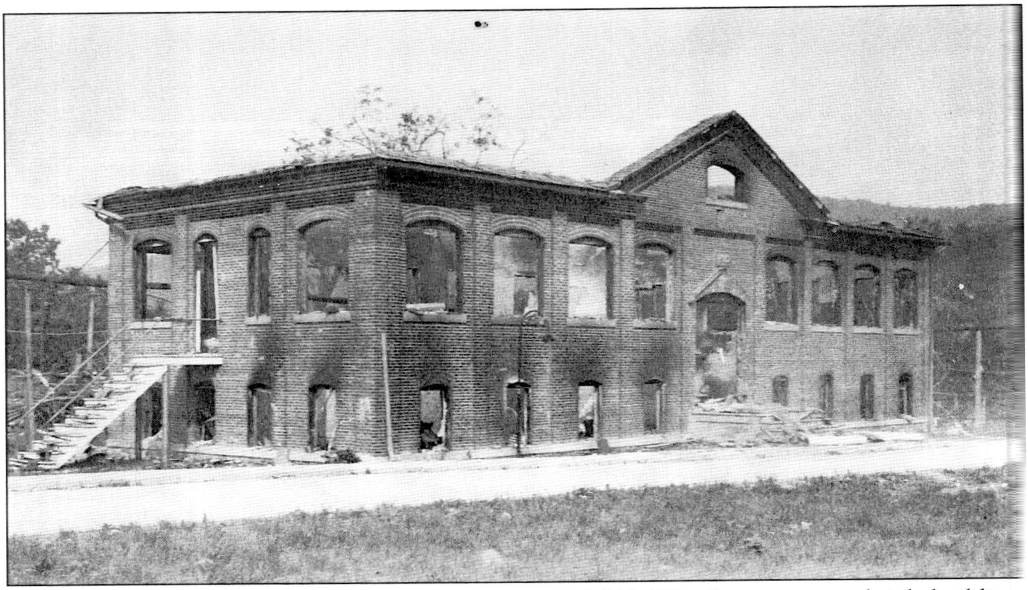

CHEMICAL LABORATORY. Built in 1884 at a cost of $33,972, this two-story brick building, 2,500 feet distant from the initial explosions, was badly exposed to the shock waves. The steel roof trusses and corrugated roof collapsed, penetrating a false ceiling. The 18-inch-thick brick walls were badly damaged. (U.S. Army Photograph-Picatinny Arsenal.)

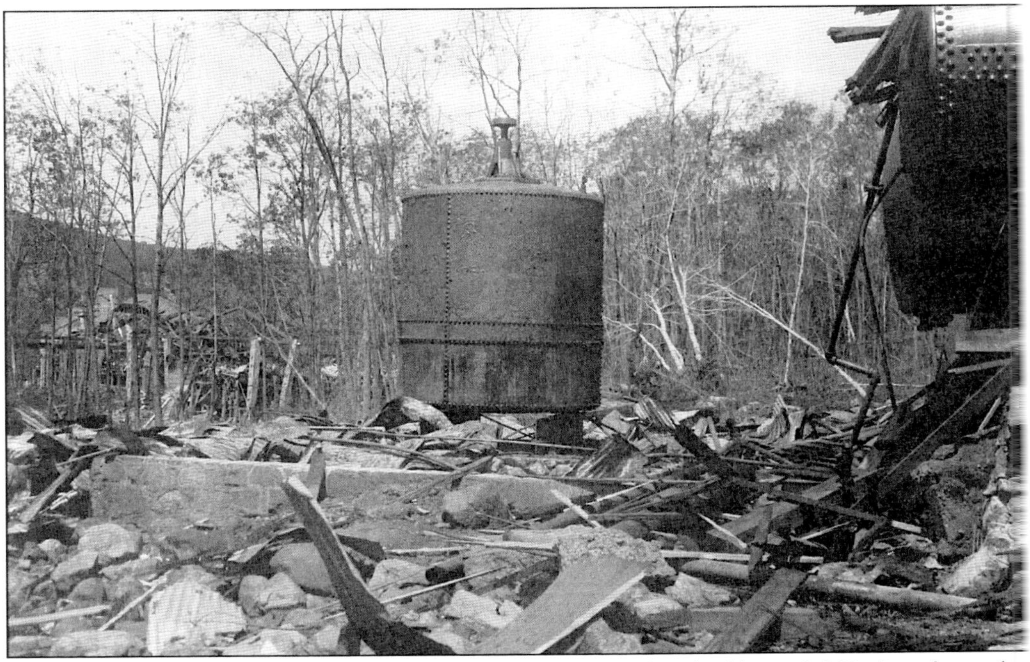

TETRYL DRY HOUSE. At the time of the explosions, this building, 2,200 feet from the explosion site, contained 500 pounds of tetryl in drying trays. The pale yellow explosive, used as a detonator, did not explode, but the tile-walled, one-story building, constructed in 1920 at a cost of $3,220, was wrecked. (U.S. Army Photograph-Picatinny Arsenal.)

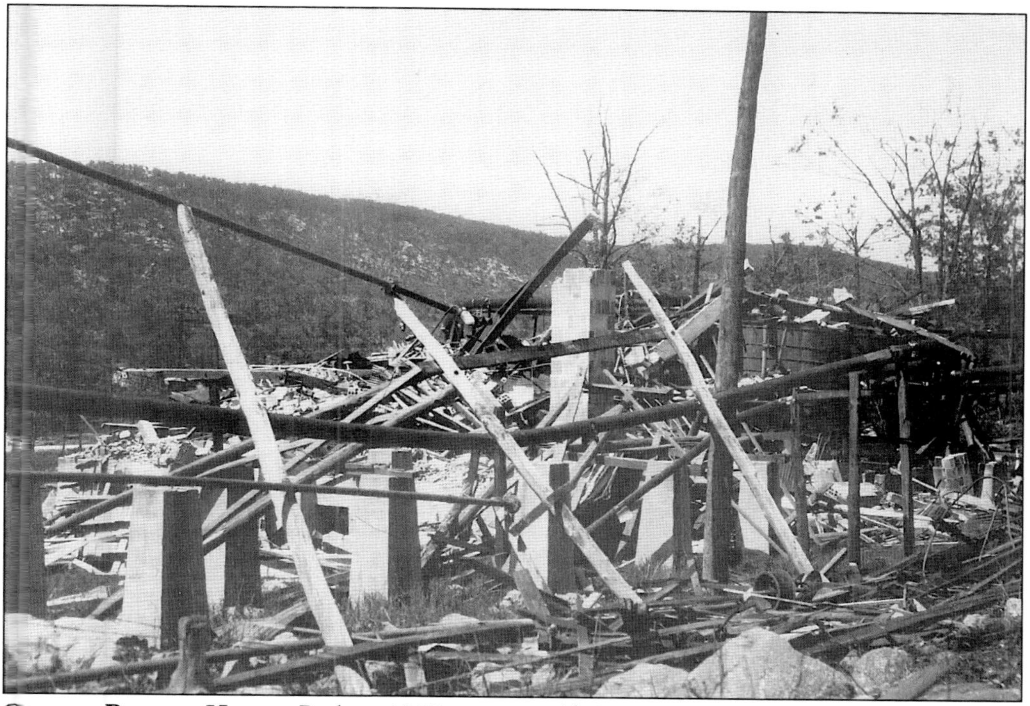

COTTON BOILING HOUSE. Built in 1907 at a cost of $17,515, this tile and steel building, 1,250 feet from the explosion site, was leveled by the initial blast and subsequent shock waves. Only concrete piers for the foundation remained intact, supporting a wooded floor littered with debris and parts of the roof. (U.S. Army Photograph-Picatinny Arsenal.)

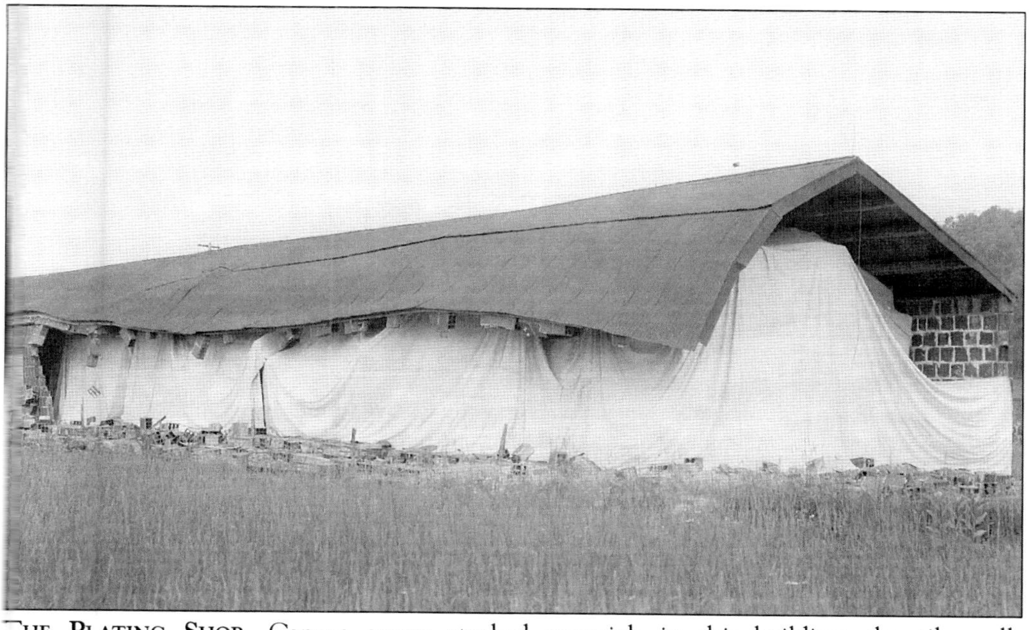

THE PLATING SHOP. Canvas covers stacked materials in this building, the tile walls of which were partially demolished on one side by the shock waves that bounced from hill to hill. This building was constructed in 1918 at a cost of $18,500. (U.S. Army Photograph-Picatinny Arsenal.)

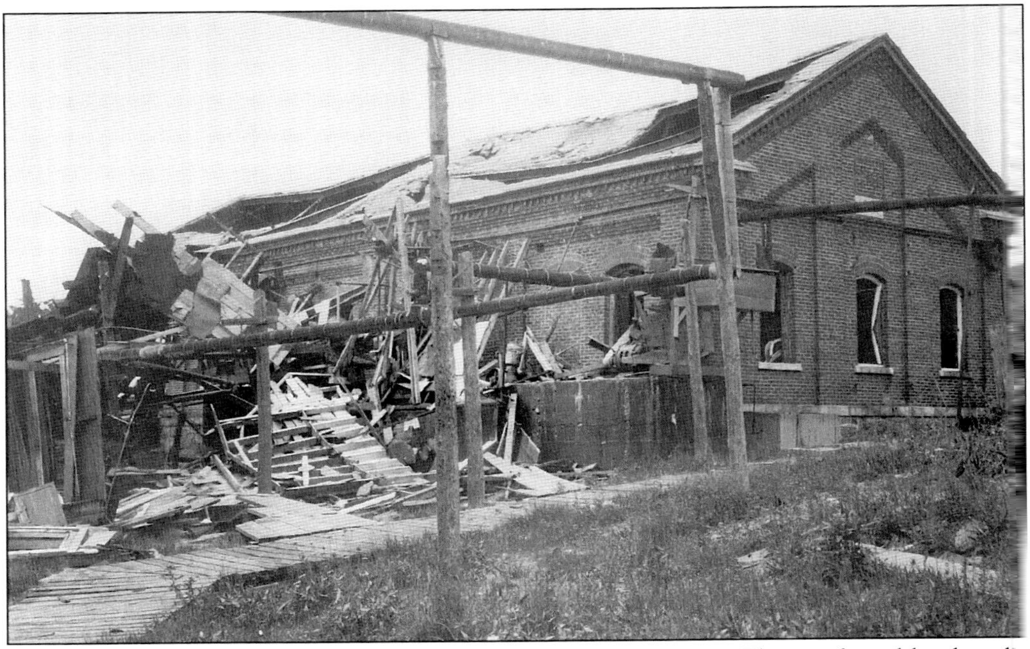

The Highly Explosive Experimental Propellant Building. The reinforced brick wal of this 1904 constructed building, at a cost of $15,390, withstood the explosions and shoc waves but not the roof, windows, and frame addition, which were demolished. (U.S. Arm Photograph-Picatinny Arsenal.)

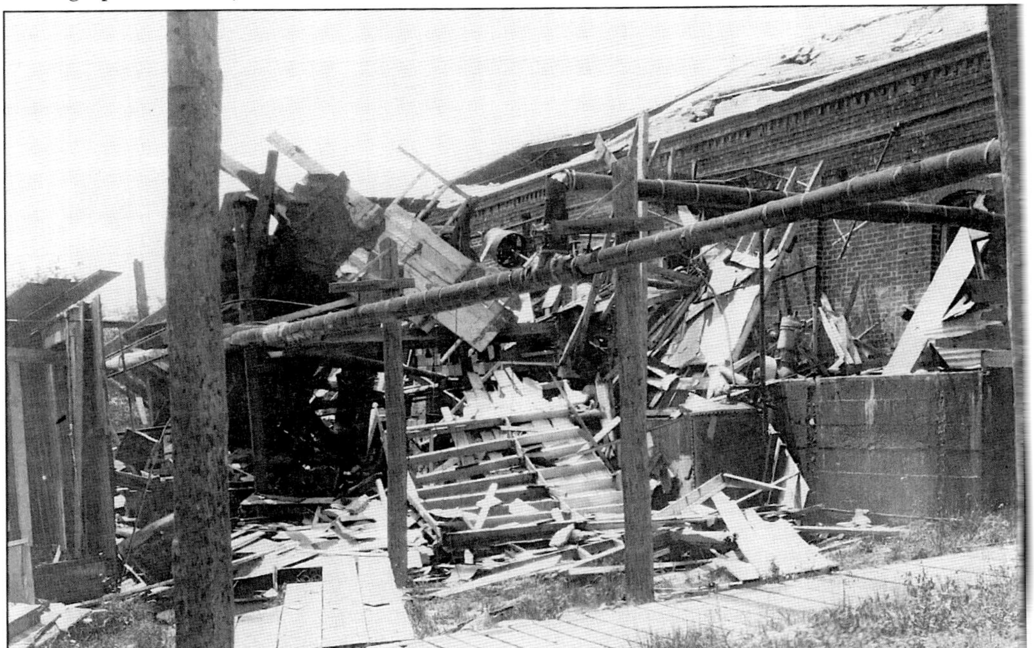

The Steamlines Survive. Standing after the explosions and shock waves were the pipes which carried steam for heat from the powerhouse to the building and the wooden telephone poles, which supported them. Surrounding them was the debris from the crushed frame addition. (U.S. Army Photograph-Picatinny Arsenal.)

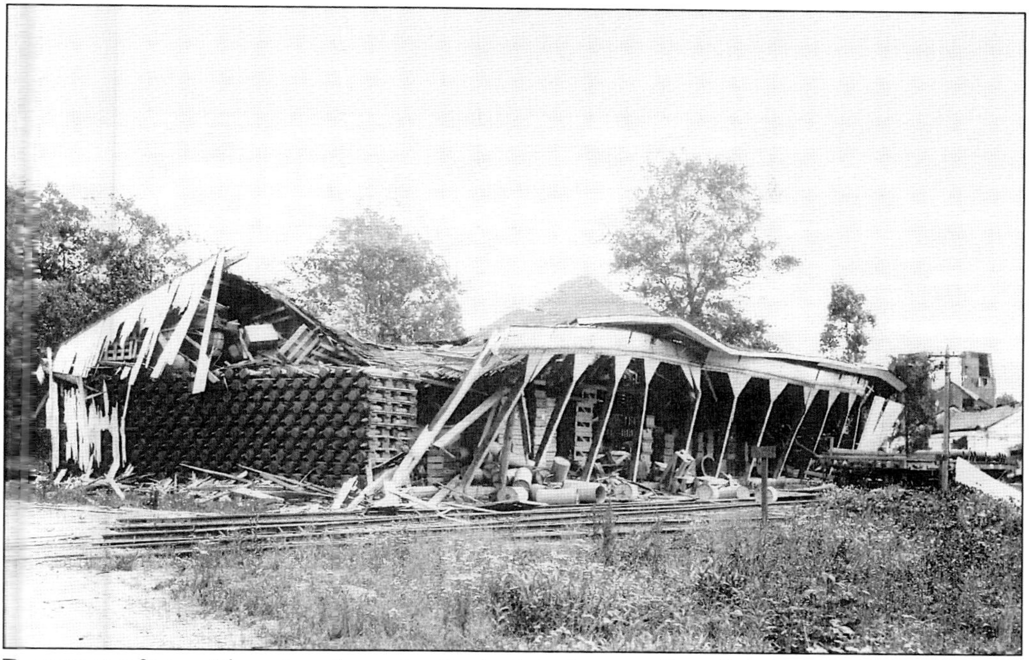

PROJECTILE SHED. The ruins of a projectile shed, its frame walls collapsed and corrugated metal roof felled by the explosions and shock waves. Sheets of metal from the corrugated roof hang over the cross arms of a telephone pole, which survived. Containers litter the railroad tracks. (U.S. Army Photograph-Picatinny Arsenal.)

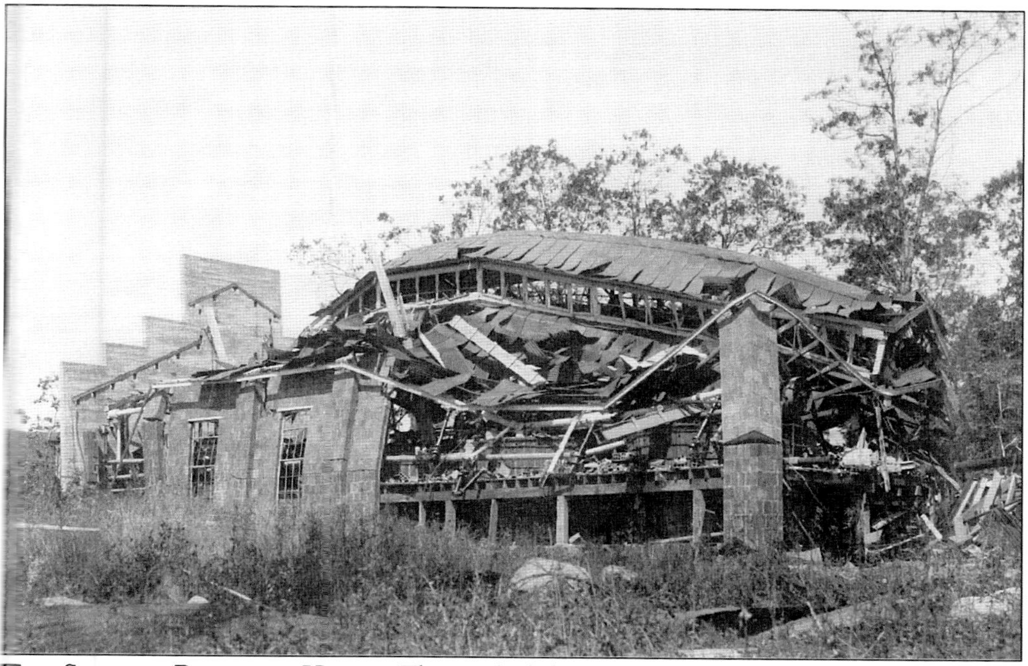

THE SOLVENT RECOVERY HOUSE. The roof of this one-story brick and concrete building, 2,000 feet from the explosion site, was shattered and sections of its brick walls cracked or demolished by the shock waves. It was built in 1909 at a cost of $23,168. (U.S. Army Photograph-Picatinny Arsenal.)

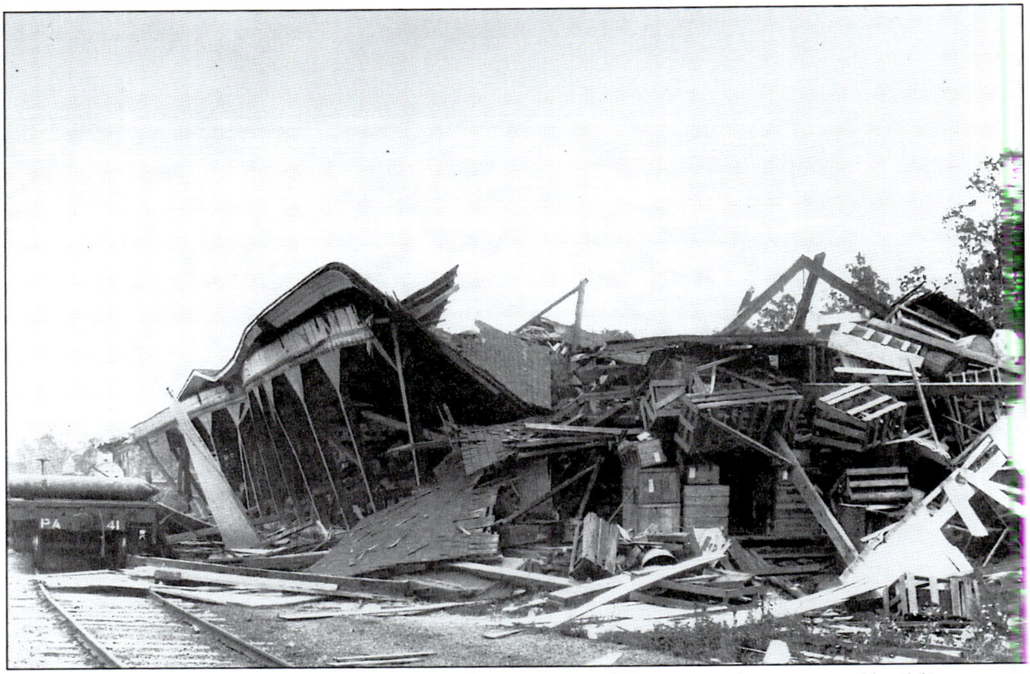

THE STORAGE SHED. The force of the explosions tossed the twisted remains of buildings onto railroad tracks, demolishing railroad cars and engines. Brick walls in some magazines were glazed by the intense heat that was created by burning munitions that did not explode. (U.S. Army Photograph-Picatinny Arsenal.)

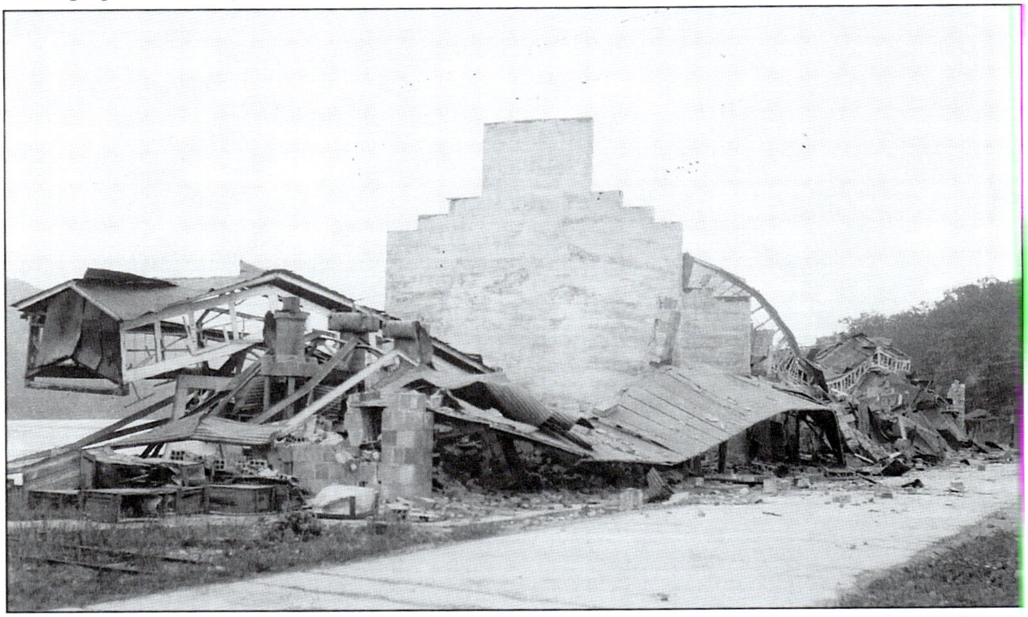

SMOKELESS POWDER OPERATION. This dehydration building, 1,900 feet from the explosion site, and three adjacent tile structures, used for smokeless powder operations, were completely wrecked despite a number of 12-inch-thick partition walls. Built in 1922 at a cost of $9,600, they were used for dehydration, mixing, and pressing. Their equipment escaped serious damage. (U.S. Army Photograph-Picatinny Arsenal.)

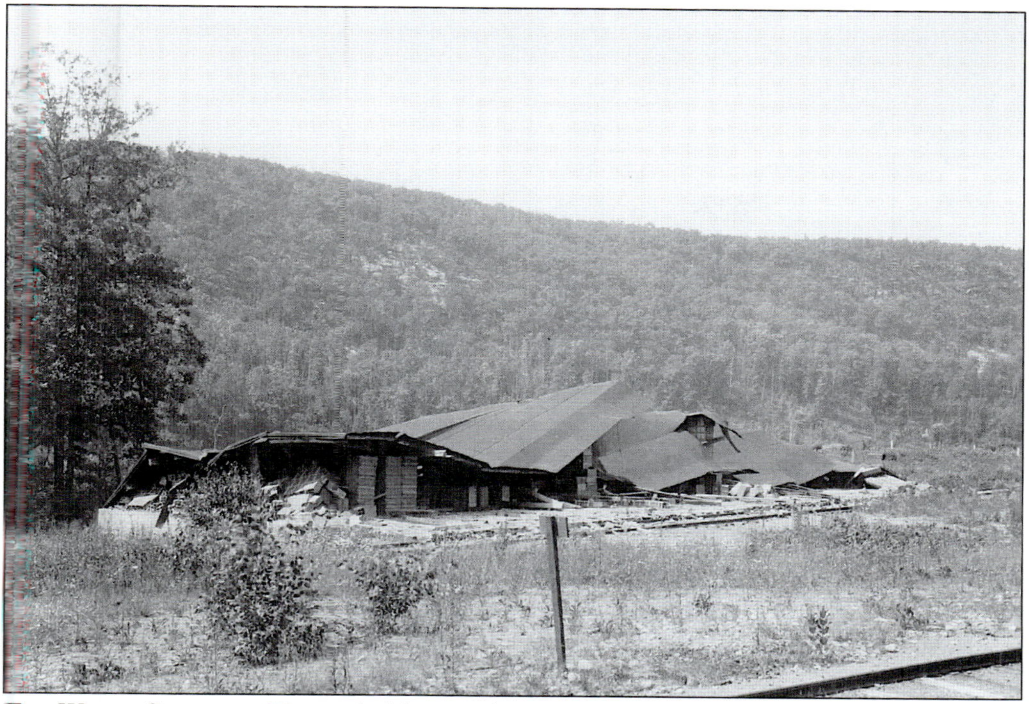

THE WALLS COLLAPSE. The roof of this storehouse caved in when the tile walls collapsed from the force of the shock waves. Three factors determined structural damage to arsenal buildings: distance from the explosion, structural strength of the buildings, and the extent to which they were screened from the explosion. (U.S. Army Photograph-Picatinny Arsenal.)

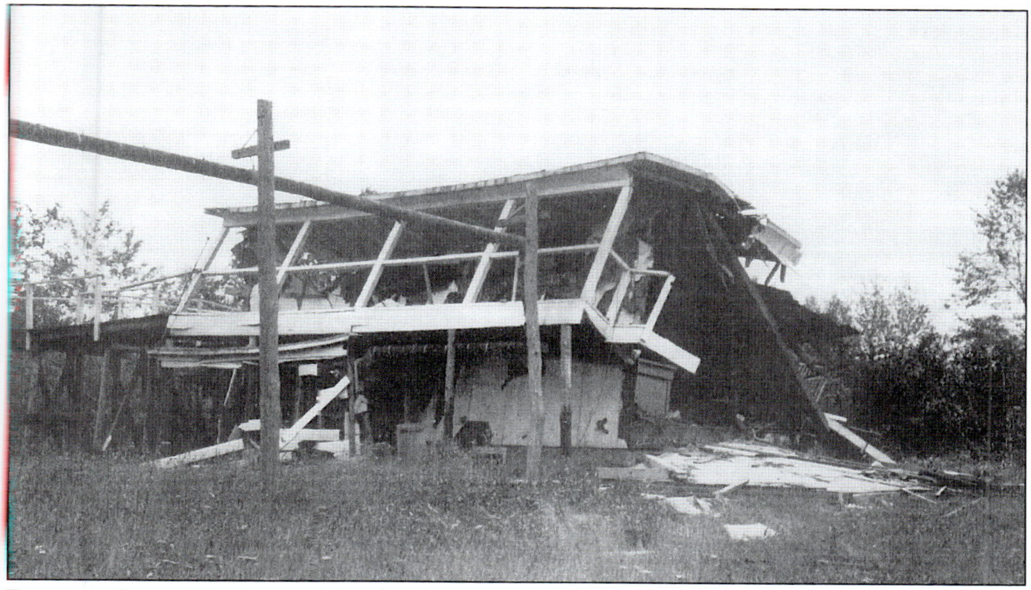

BOWLED OVER. The force of the shock waves completely demolished the walls on one side of this two-story building, bowling it over on its side. Not damaged were the steam pipes or poles supporting them. (U.S. Army Photograph-Picatinny Arsenal.)

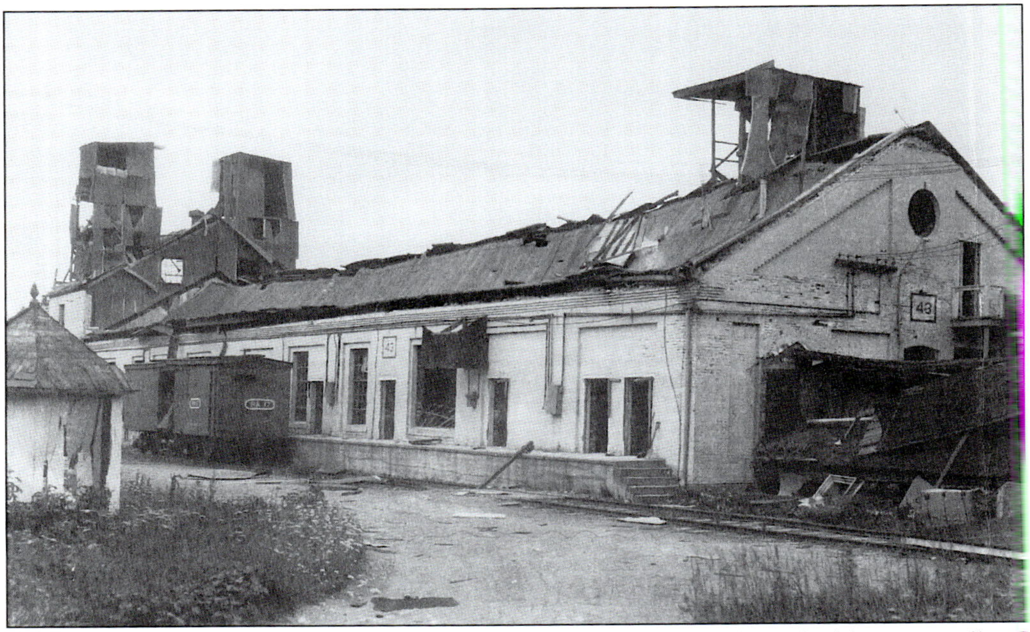

Bag Loading Plant. The brick walls on this building 2,500 feet from the blast site, suffered little damage, but the wooden roof and frame superstructure partially collapsed, their debris falling on the railroad tracks and a boxcar. The frame storage buildings adjacent to the plant were demolished. (U.S. Army Photograph-Picatinny Arsenal.)

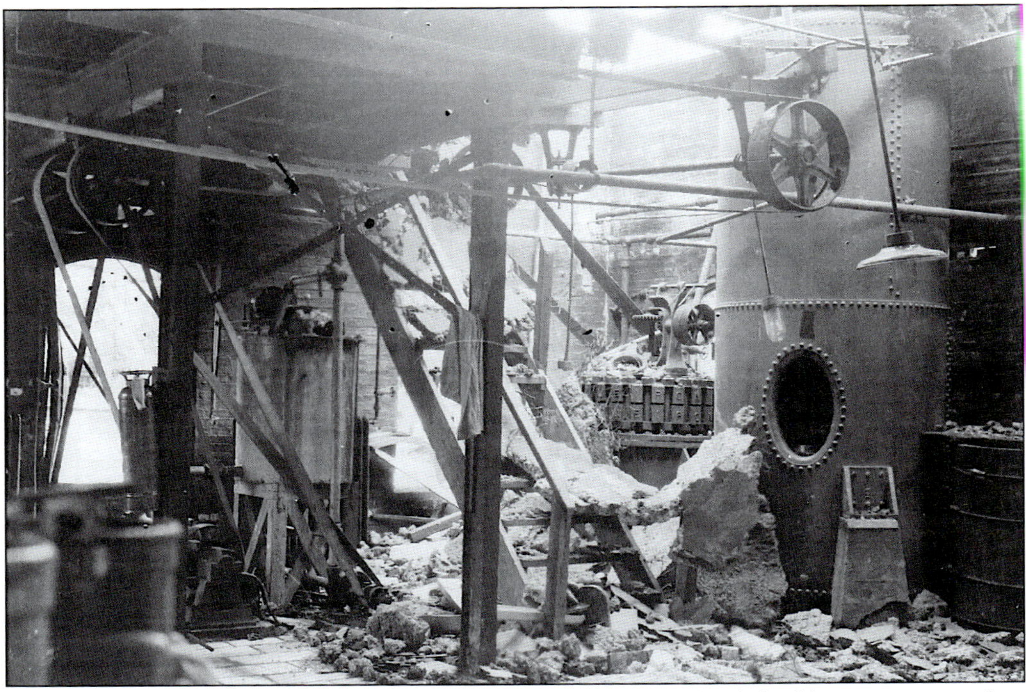

Experimental Propellant Building. The interior of this brick building, constructed in 1904 at a cost of $15,390, was littered with debris from the cement roof that collapsed. The effects of the explosion showed that the peaked-type roof was the weakest part of a building. (U.S. Army Photograph-Picatinny Arsenal.)

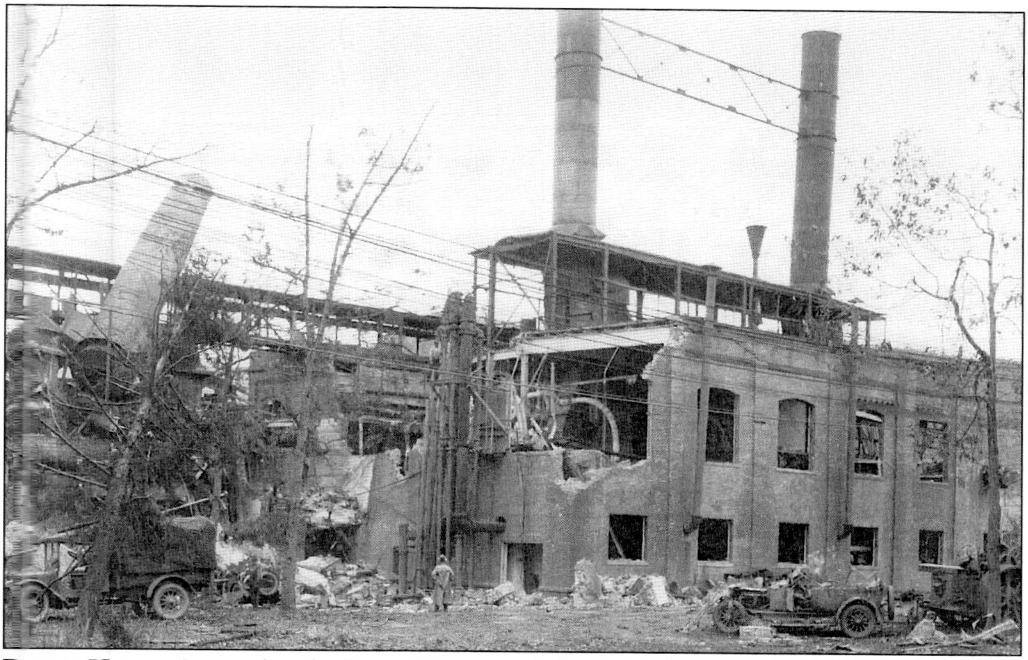

POWER HOUSE. Located at the foot of Picatinny Lake, 1,500 feet from the blast site, the north and south ends of this brick building constructed with interior steel columns and steel support floors, plus one stack, were demolished. The center section, two stacks, the coal conveying systems, and ash skip hoist were undamaged. (U.S. Army Photograph-Picatinny Arsenal.)

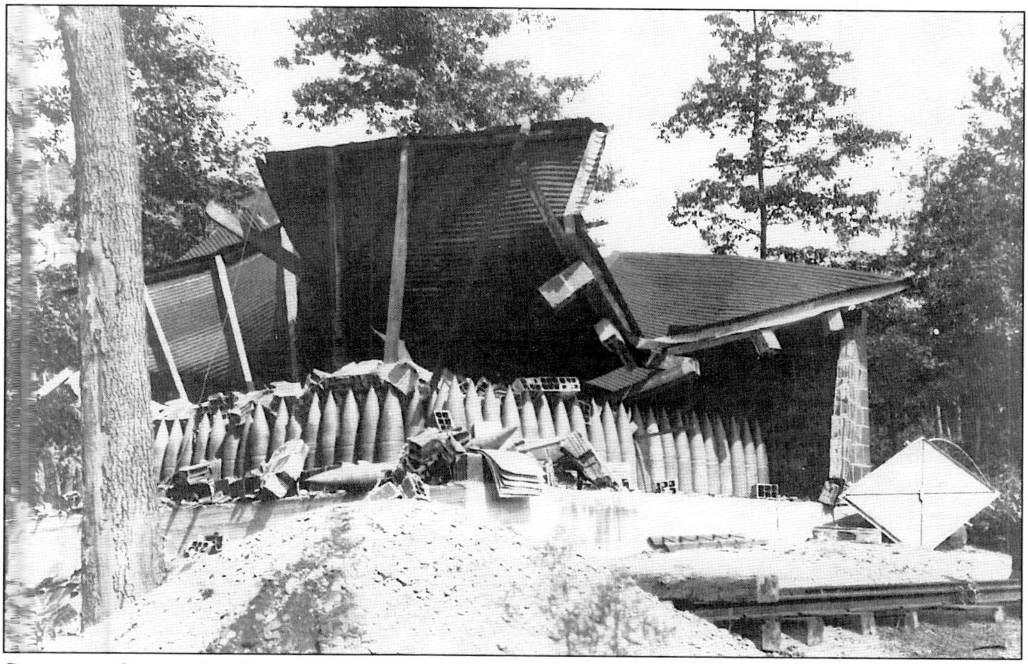

STORAGE SHED. Artillery shells are stacked in rows in this projectile storage shed, the tile walls and roof of which collapsed. Dotted throughout the arsenal, the sheds were part of the 5-3 buildings interconnected by railroad tracks, roads, paths, and steam lines. (U.S. Army Photograph-Picatinny Arsenal.)

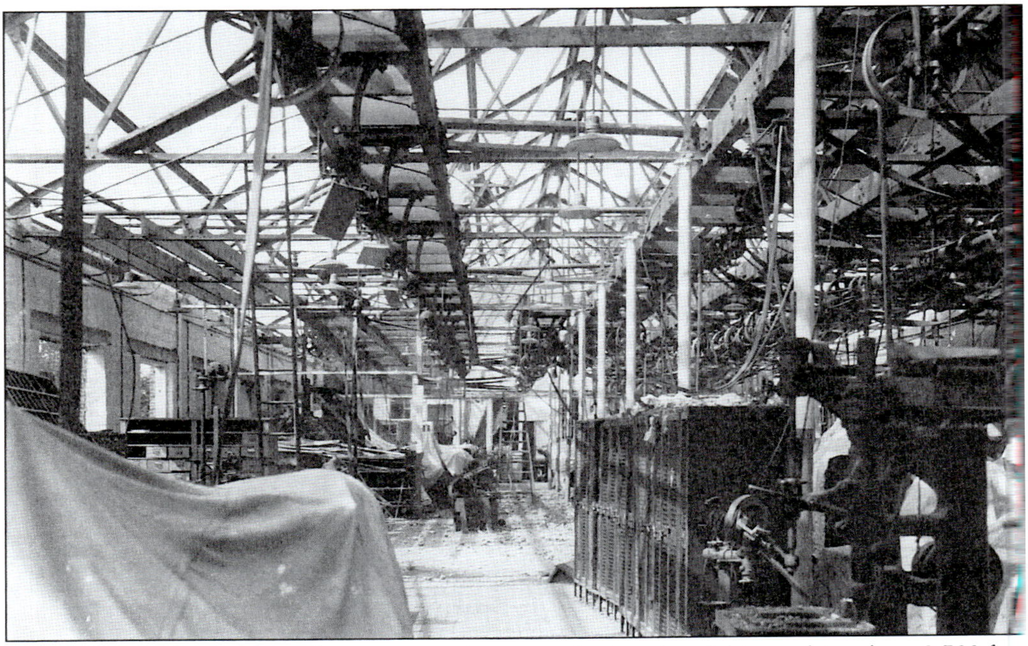

THE MACHINE SHOP. Tarpaulins cover machinery in the interior of this machine shop, 2,700 feet from the blast site. The 18-inch-thick brick walls were undamaged but the 4-inch-thick concrete roof slab, covered with slate, disintegrated from the force of the shock waves. The building was constructed in 1904 at a cost of $28,500. (U.S. Army Photograph-Picatinny Arsenal.)

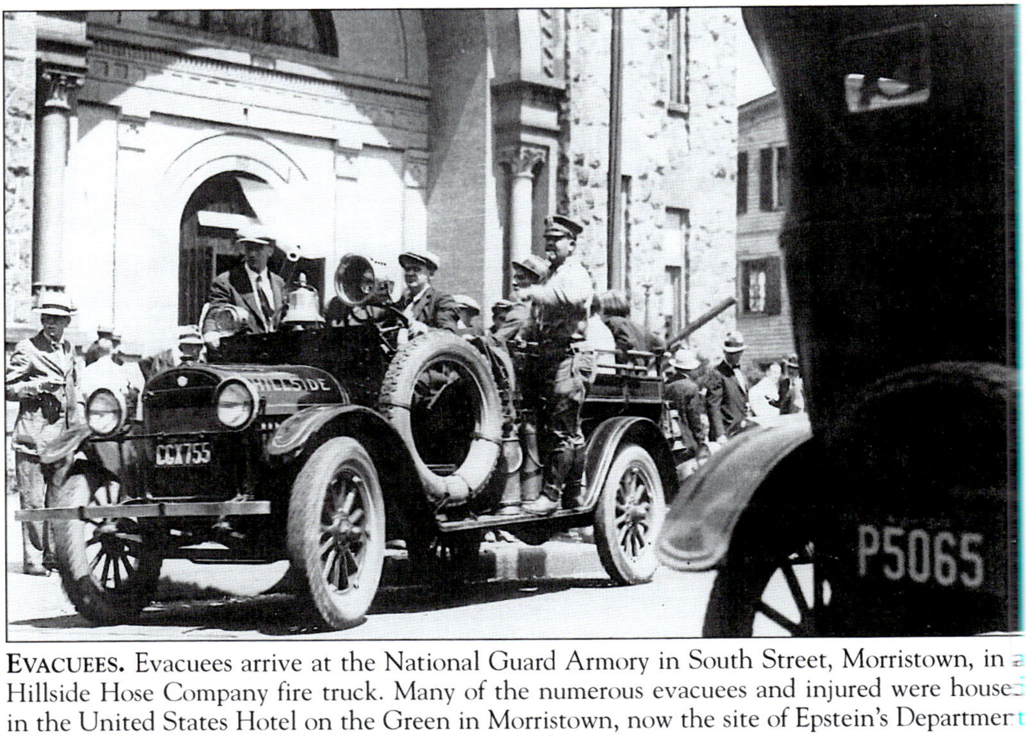

EVACUEES. Evacuees arrive at the National Guard Armory in South Street, Morristown, in a Hillside Hose Company fire truck. Many of the numerous evacuees and injured were housed in the United States Hotel on the Green in Morristown, now the site of Epstein's Department Store. (U.S. Army Photograph-Picatinny Arsenal.)

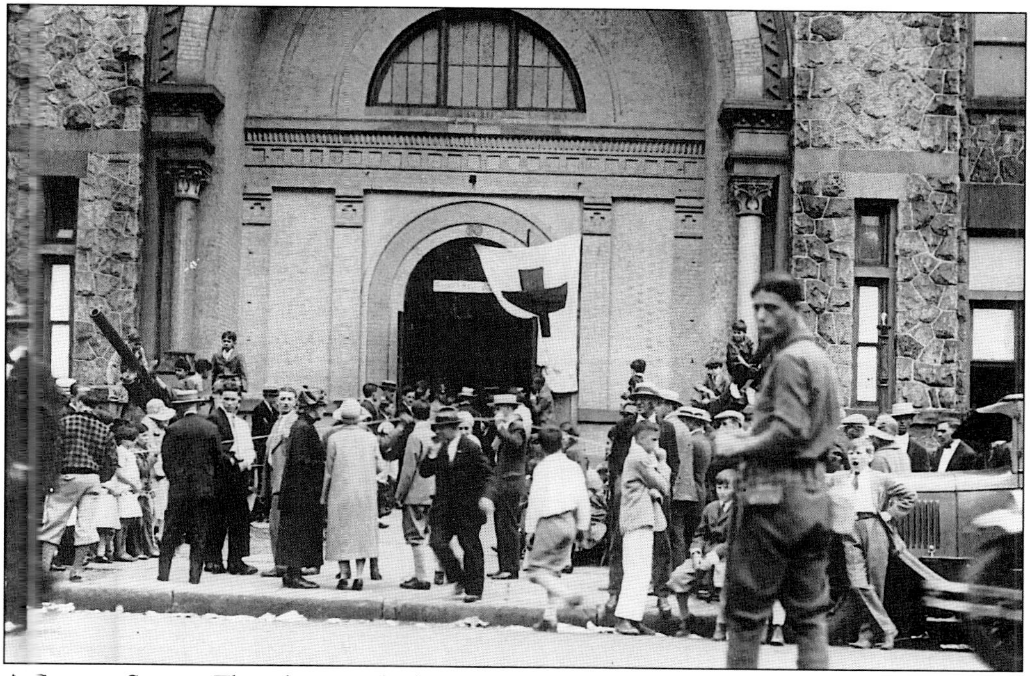

A Street Scene. This photograph shows South Street in Morristown, outside the National Guard Armory, where scores gathered to await the arrival of evacuees and lend assistance. Physicians and nurses from a wide area responded to the call for assistance. (U.S. Army Photograph-Picatinny Arsenal.)

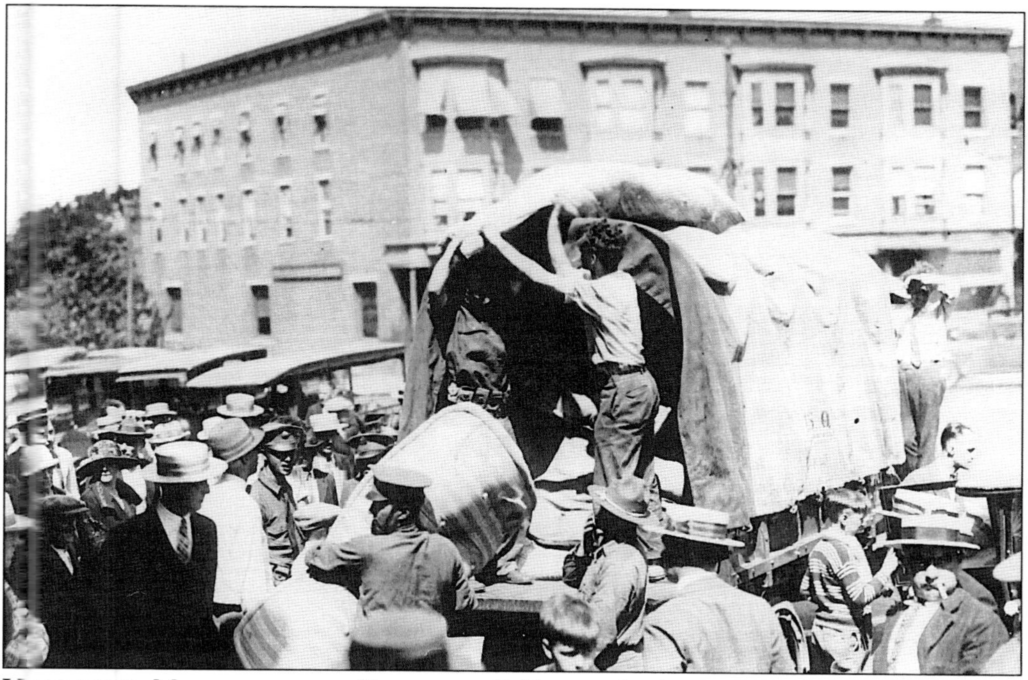

Unloading Mattresses for Evacuees. Soldiers unload mattresses from an army truck outside the United States Hotel on the Green, in Morristown. The evacuees came from a wide area surrounding the arsenal. (U.S. Army Photograph-Picatinny Arsenal.)

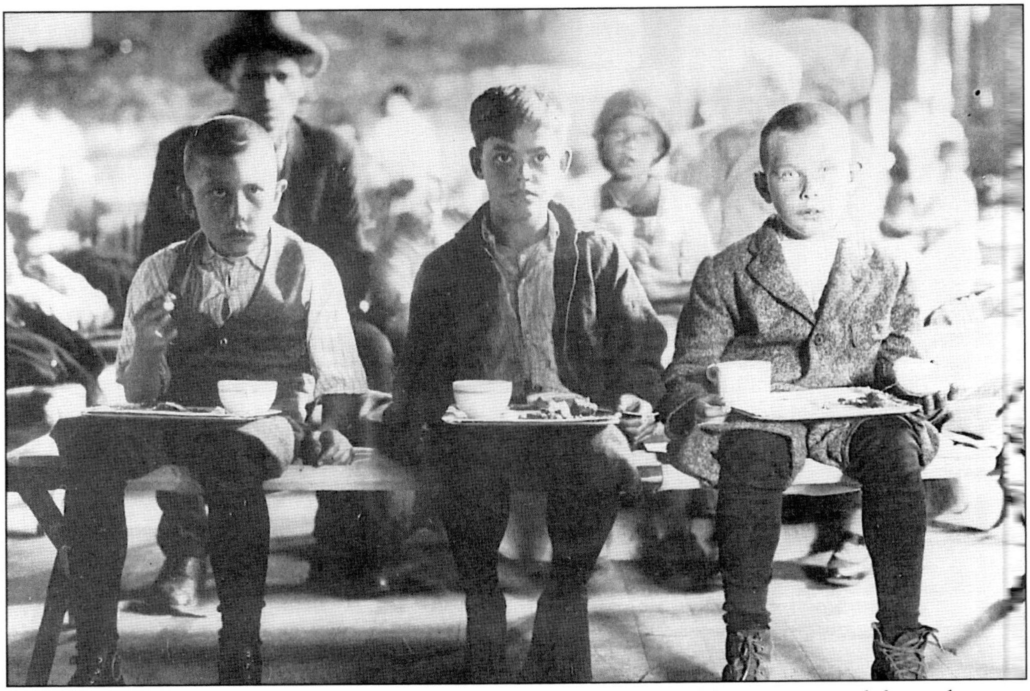

THEIR FIRST MEAL IN MANY HOURS. Three of the many children evacuated from the area around the arsenal enjoy a meal at the National Guard Armory on South Street, Morristown. The food was provided by the Red Cross, area churches, and restaurants. (U.S. Army Photograph-Picatinny Arsenal.)

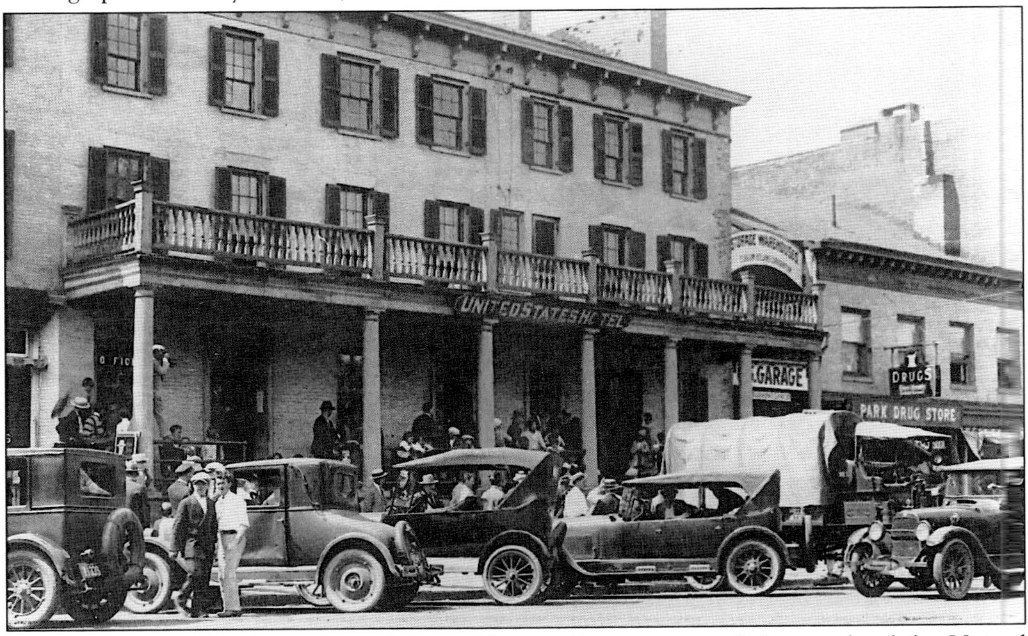

A SCENE OUTSIDE THE UNITED STATES HOTEL. Evacuees crowd the porch of the United States Hotel on the Green in Morristown, two blocks from the National Guard Armory. Cars, many of which brought evacuees to Morristown, line the curb. In the background is the army truck from which mattresses were unloaded. (U.S. Army Photograph-Picatinny Arsenal.)

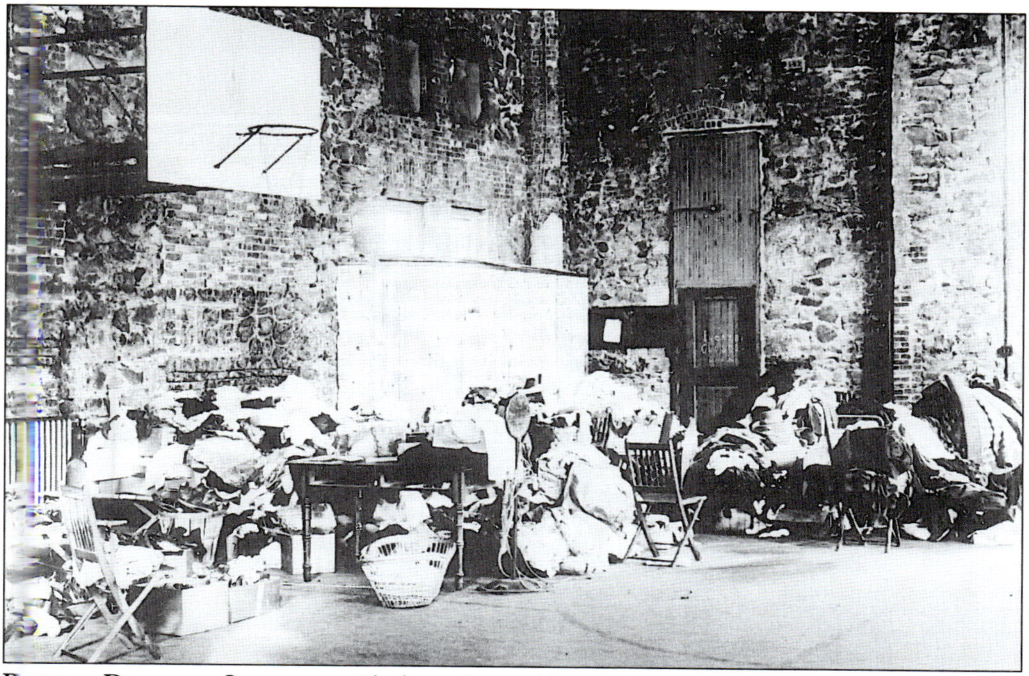

PILES OF DONATED CLOTHING. Clothing donated by Morris County residents for the evacuees sits on tables on the basketball court of the National Guard Armory on South Street, Morristown. (U.S. Army Photograph-Picatinny Arsenal.)

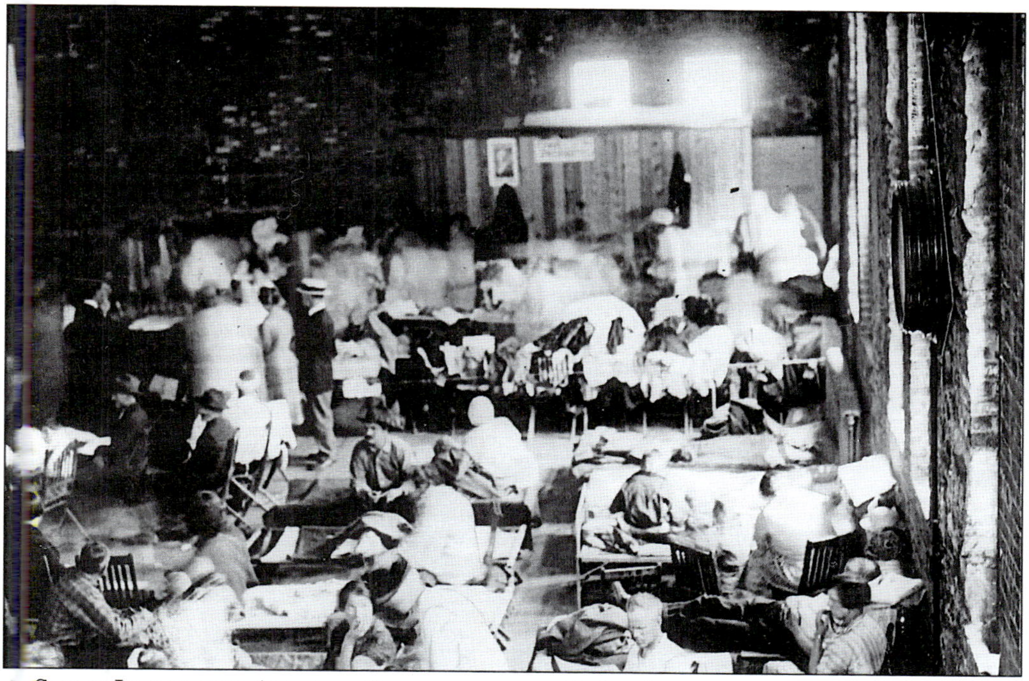

A SCENE INSIDE THE ARMORY. Evacuees, some with bandages covering wounds, sit on cots set up inside the South Street National Guard Armory. The armory, construction of which was financed by two area millionaires, was on the site of the burned-out Library Lyceum. (U.S. Army Photograph-Picatinny Arsenal.)

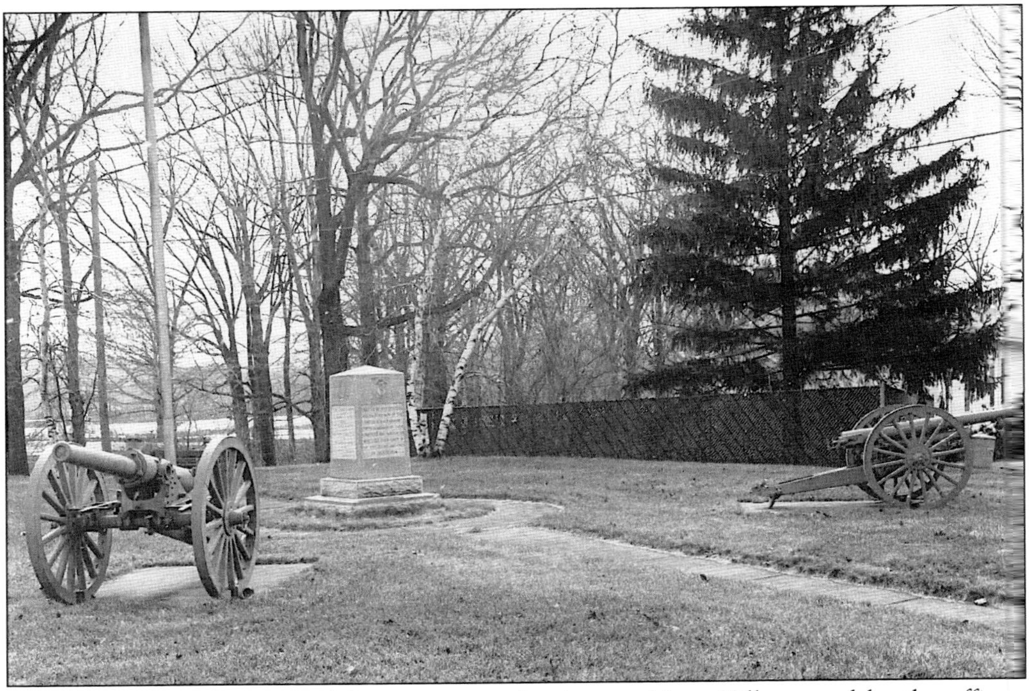

MARINE CORPS MEMORIAL. This granite marker is atop Navy Hill, erected by the officers and enlisted men of the U.S. Marine Corps in memory of their comrades killed in the 1926 explosion. Most of the dead were in the initial firefighting party. All were within 300-400 feet of the explosion. (U.S. Army Photograph-Picatinny Arsenal.)

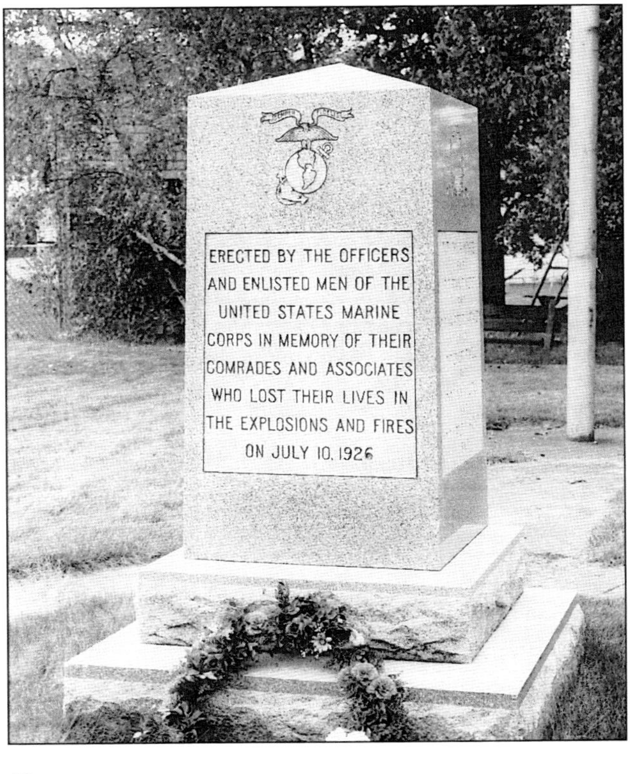

GRANITE MARKER. Flanked by two naval cannon and a towering flag staff, the marker erected shortly after the explosion honors the 11 Marines, 1 enlisted man, and 4 commissioned officers killed and 38 officers and privates injured in the 1926 explosion. (U.S. Army Photograph-Picatinny Arsenal.)

MAJ. NORMAN F. RAMSEY. Commanding officer at the time of the explosion, Ramsey was awarded the Soldier's Medal for heroism for leading a party of eight men into the burning area, despite exploding shells and small arms ammunition, to recover the body of Lt. George W. Bott Jr. (U.S. Army Photograph-Picatinny Arsenal.)

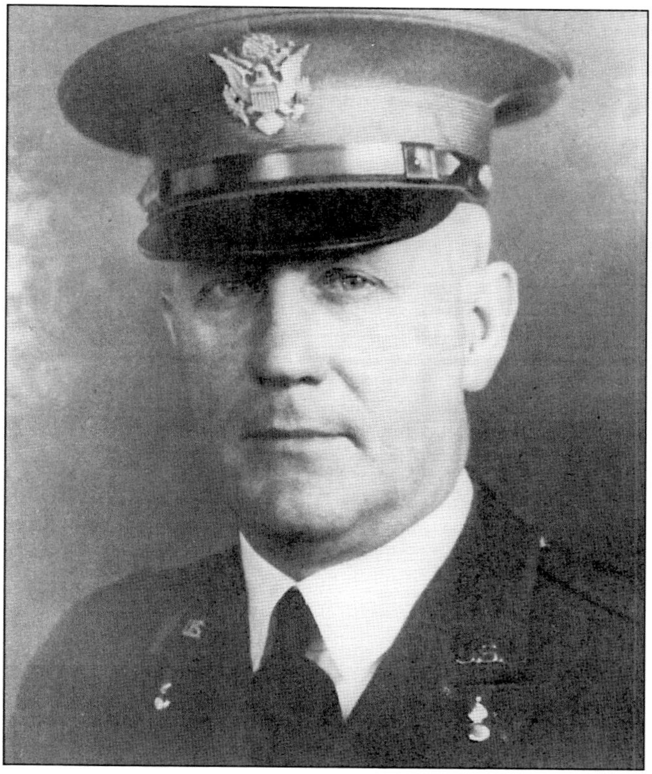

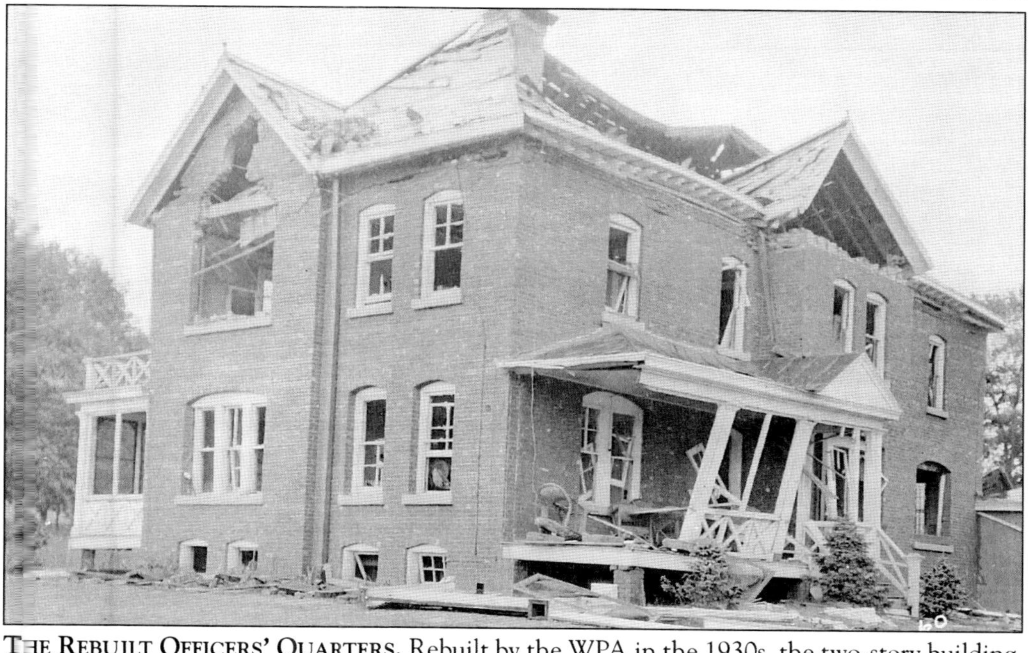

THE REBUILT OFFICERS' QUARTERS. Rebuilt by the WPA in the 1930s, the two-story building, built in 1884 at a cost of $14,000, has served as officers quarters, a fire station, guard house, and schoolhouse. Two shades of red brick are visible: the old brick with which the house was constructed and a new brighter colored red brick used by the WPA masons. (U.S. Army Photograph-Picatinny Arsenal.)

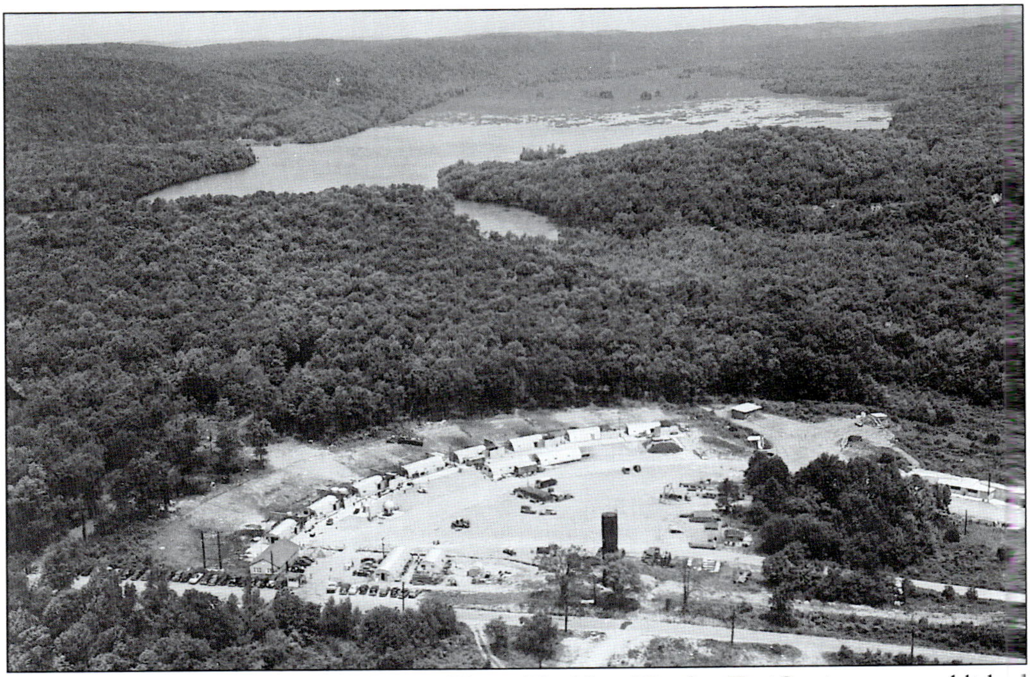

NAVAL ROCKET TEST STATION, AERIAL VIEW. The Naval Rocket Test Station was established at Lake Denmark after four men, the founders of Reaction Motors, Inc., were asked to stop test firing liquid-fueled rockets in a sandpit in Pompton Plains, New Jersey, by the town administration. (U.S. Army Photograph-Picatinny Arsenal.)

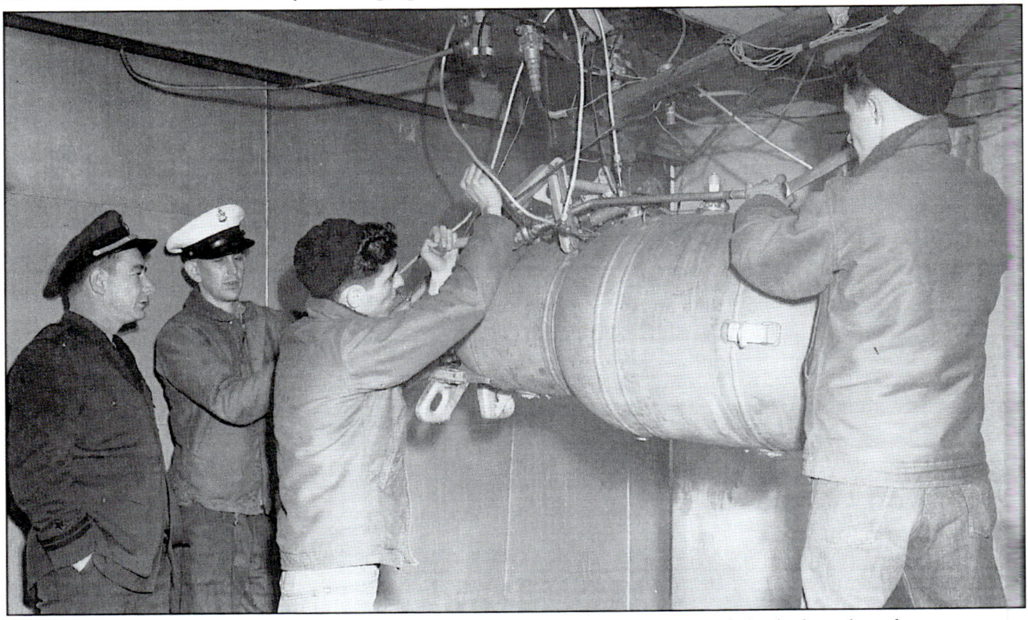

ROCKET PREPARATION. Workmen and technicians prepare a liquid-fueled rocket for testing at the Naval Rocket Test Station, established at Lake Denmark on the site of the powder depot which exploded in 1926. Accomplishments at the site included development of engines for the X-1, the first rocket to break the sound barrier, and the X-15, which reached a speed of 1,500 miles an hour in 1960. (U.S. Army Photograph-Picatinny Arsenal.)

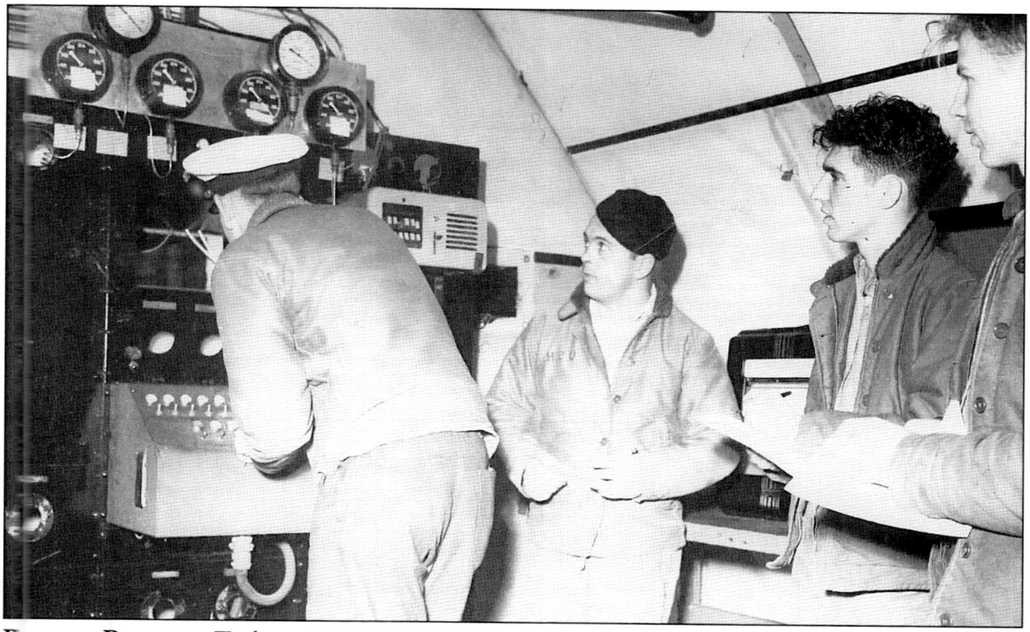

RECORD RESULTS. Technicians prepare to test fire a liquid-fueled rocket at the Naval Rocket Test Station at Lake Denmark while other technicians record data. The station and testing facilities were closed in 1960. A plaque in front of the Parker Building states, "Picatinny Arsenal, Lake Denmark Area, a Gateway to Space." (U.S. Army Photograph-Picatinny Arsenal.)

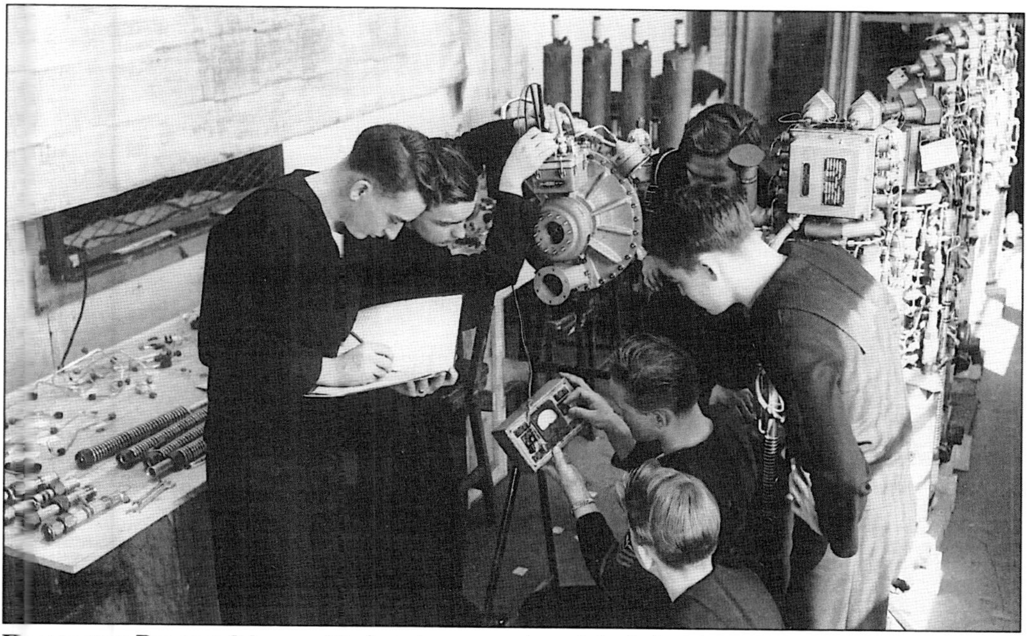

EXAMINING ROCKET MOTOR. Technicians test a liquid-fueled rocket motor at the Naval Rocket Test Station at Lake Denmark. Different parts of the motor are on the rack in the background and on the work bench. This work helped develop the XLR II rocket engine, which powered Capt. Chuck Yeager's supersonic flight in the Bell X-1 airplane on October 14, 1947. (U.S. Army Photograph-Picatinny Arsenal.)

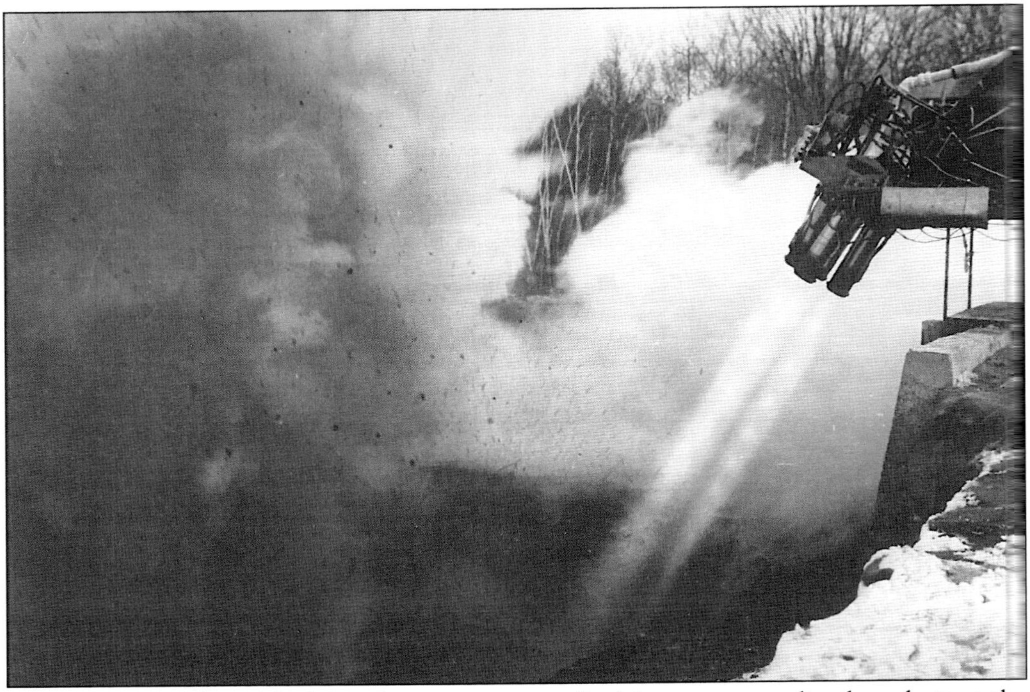

TEST FIRING. A liquid-fueled rocket motor is test fired into a cement-lined sand pit at the Naval Rocket Test Station at Lake Denmark. The force of the blast sent a shower of sand, dirt, and smoke into the air. The XLRs, worked on at Picatinny, powered the first 24 flights of the record-setting X-15 rocket plane and a family of lifting bodies, whose flight data influenced the design of the space shuttle. (U.S. Army Photograph-Picatinny Arsenal.)

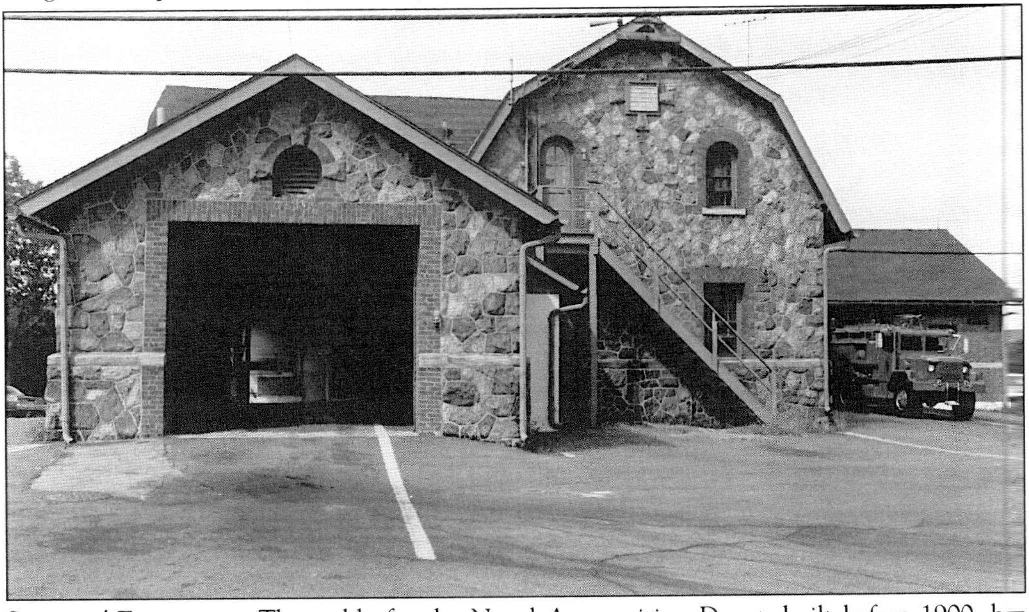

STABLE / FIREHOUSE. The stable for the Naval Ammunition Depot, built before 1900, has been converted into an arsenal firehouse. Its thick stone walls of blue puddingstone were not damaged by the 1926 explosion. The brick addition was added at a later date. (U.S. Army Photograph-Picatinny Arsenal.)

Three

WPA Rehabilitation 1932–1938

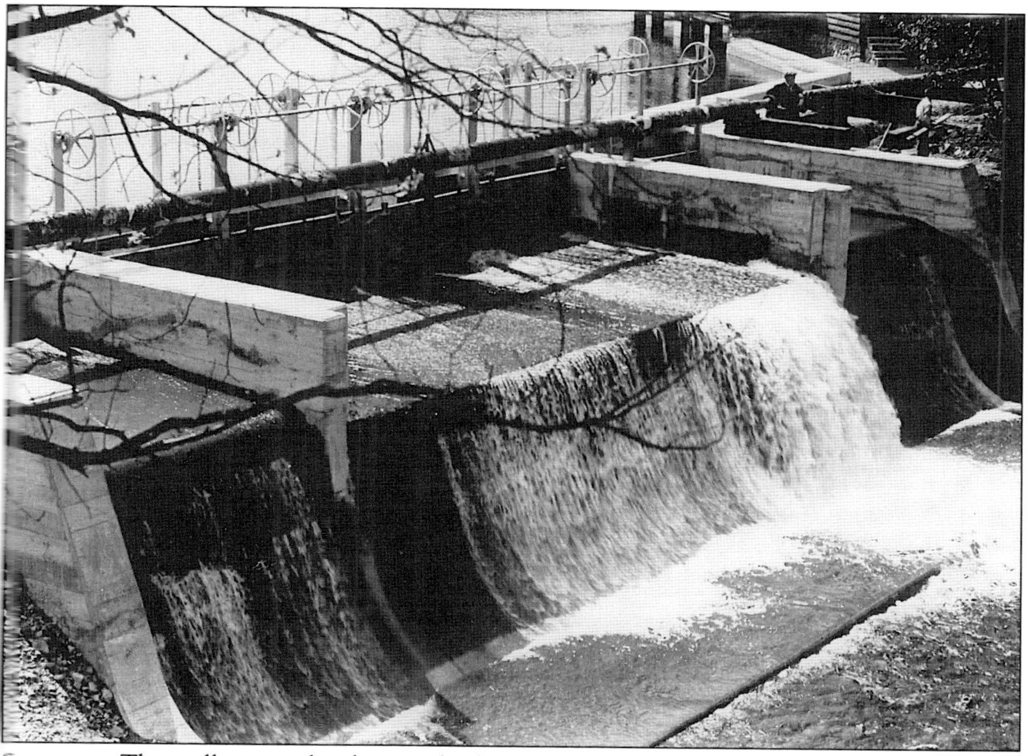

SPILLWAY. The spillway at the dam at the lower end of Picatinny Lake was rebuilt by WPA workmen. The lake, fed by converging streams from Green Pond and Lake Denmark, contains 52,500,000 gallons of water, 3,000 of which flow over the spillway each minute. (U.S. Army Photograph-Picatinny Arsenal.)

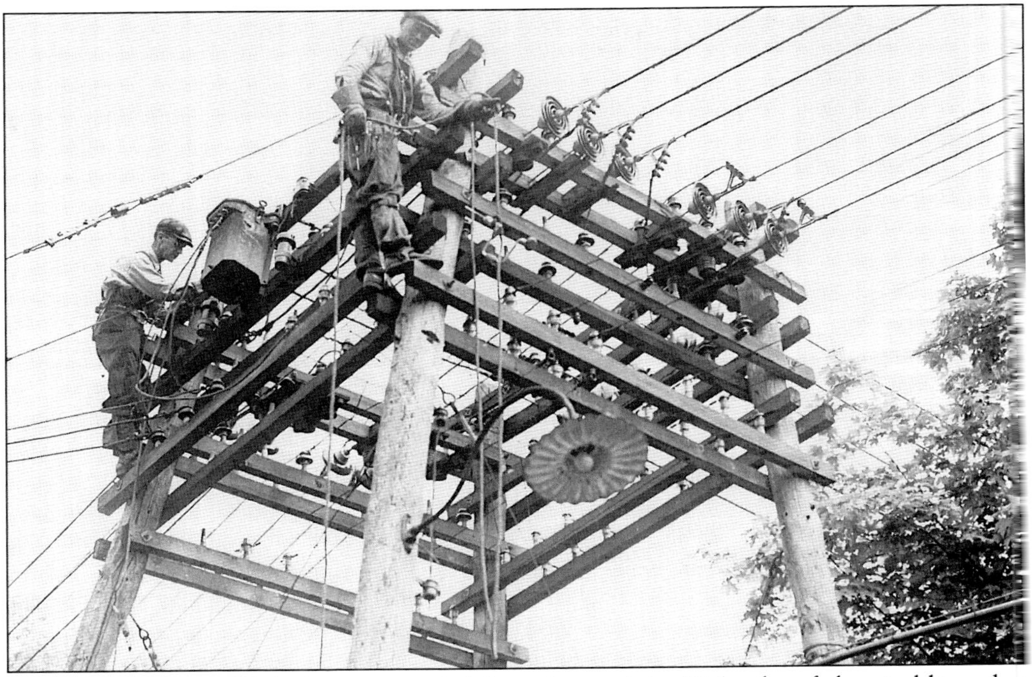

Electrical Lines. WPA electricians rehabilitated part of the 12.5 miles of electrical lines that connected the arsenal's 431 main buildings in the 1930s. In some buildings, notably officer quarters, electricity was installed in the residences. (U.S. Army Photograph-Picatinny Arsenal.)

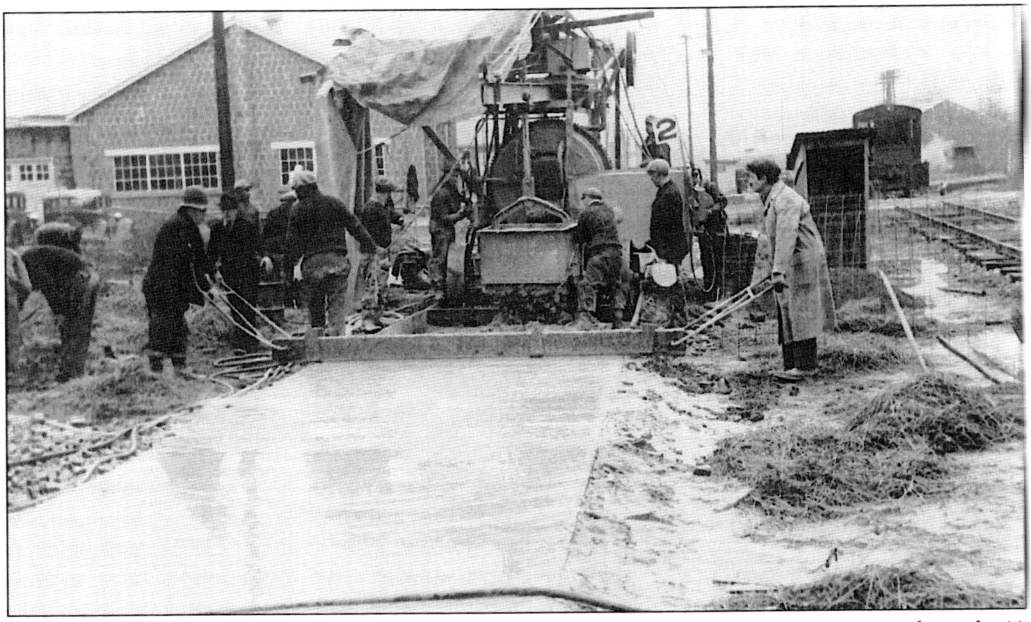

New Roads. WPA workmen constructed 44,551 linear feet of new concrete roads with 41 culverts connecting major buildings, magazines, laboratories, powder factories, and loading lines. In addition, 6,700 linear feet of macadam road was constructed, seven steel girder and concrete bridges built, and existing roadways repaired. (U.S. Army Photograph-Picatinny Arsenal.)

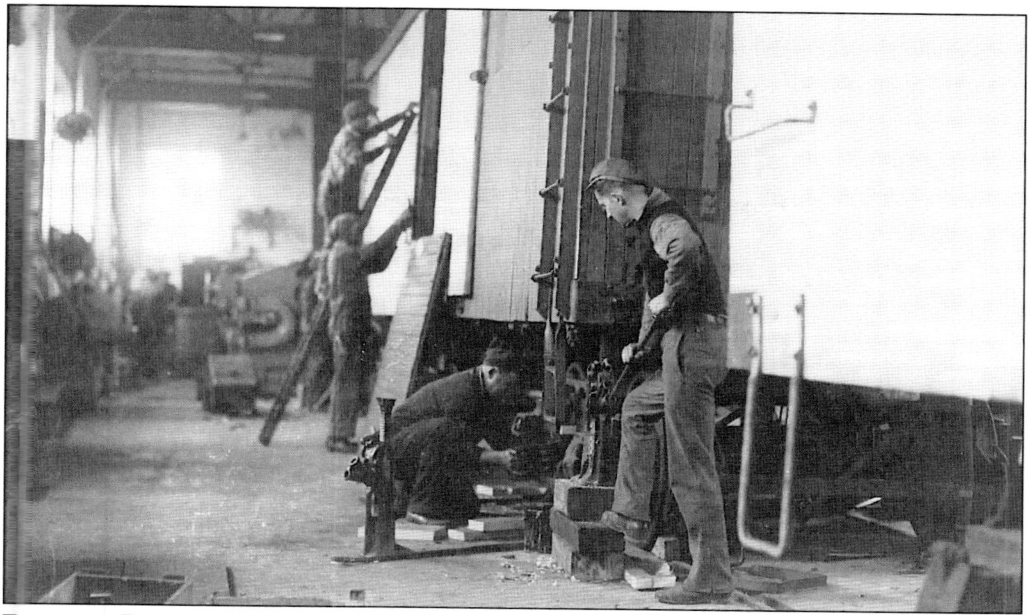

FROM THE BOTTOM UP. In many instances, WPA workers rebuilt entire railroad cars, installing new heavy beam bottoms and reinforced floors over the wheels and couplings. New sides and roofs were added and painted. In addition, workmen overhauled railroad scales, constructed [illegible],390 linear feet of new rail lines, and rehabilitated 18 miles of existing rail lines. (U.S. Army Photograph-Picatinny Arsenal.)

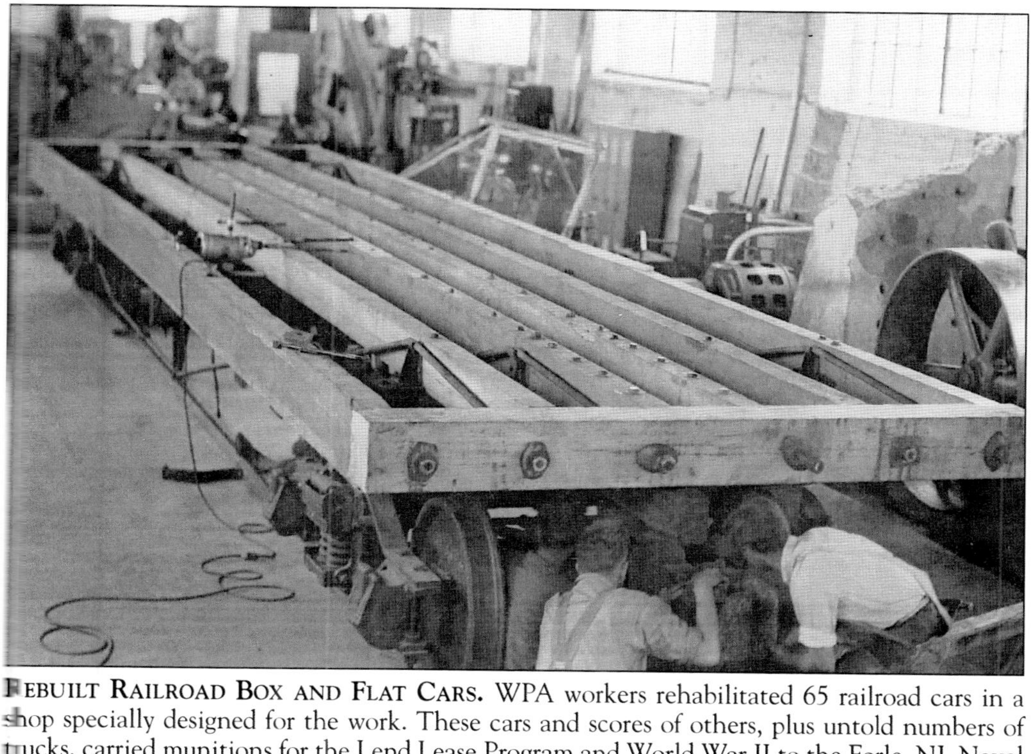

REBUILT RAILROAD BOX AND FLAT CARS. WPA workers rehabilitated 65 railroad cars in a shop specially designed for the work. These cars and scores of others, plus untold numbers of trucks, carried munitions for the Lend Lease Program and World War II to the Earle, NJ, Naval Depot where they were loaded on ships. (U.S. Army Photograph-Picatinny Arsenal.)

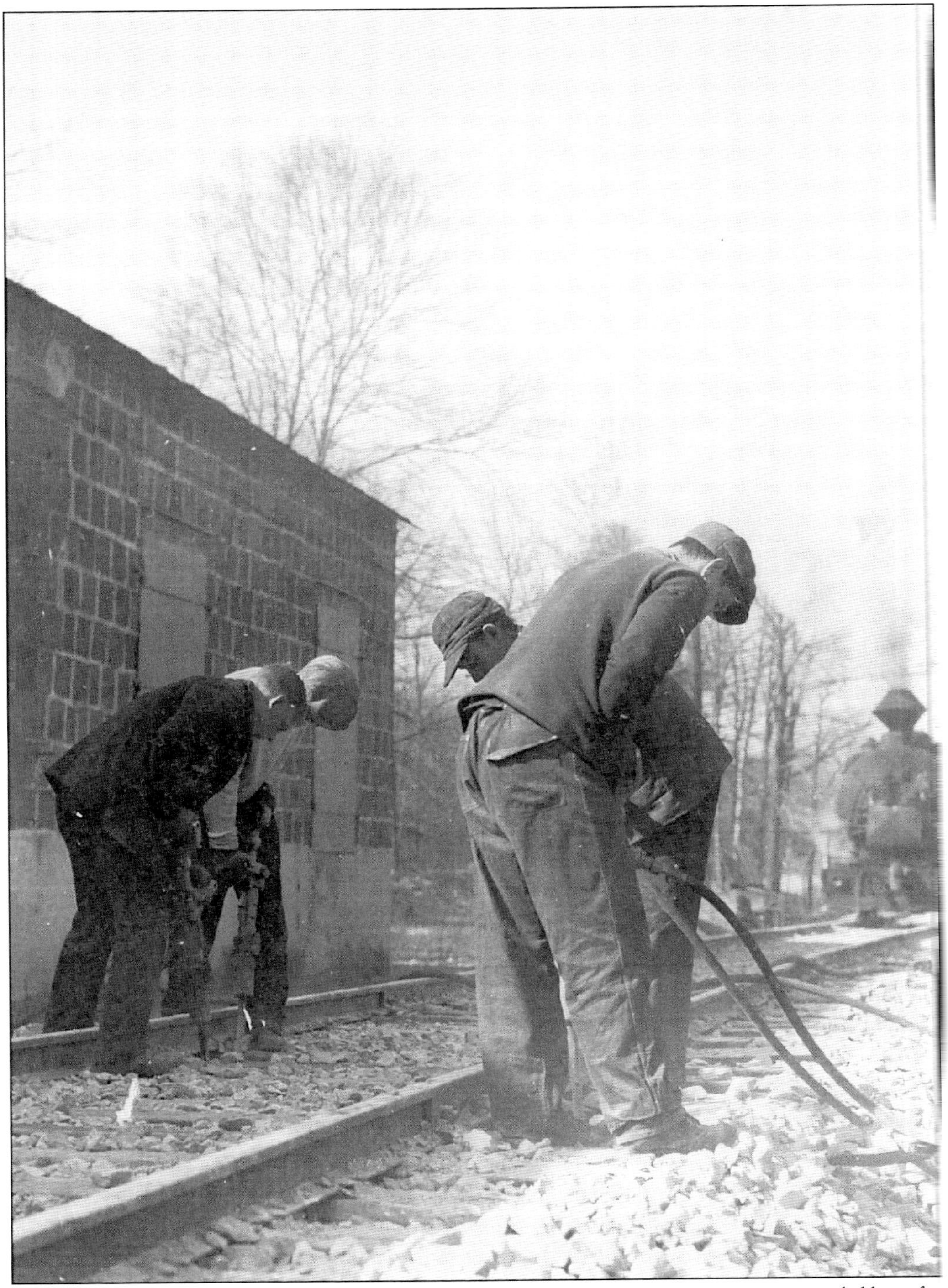

TAMPING TRACK BALLAST. WPA workmen used compressed air tampers to ensure a solid base for the new and rehabilitated railroad tracks that replaced 22,850 linear feet of outdated and unsafe track. The salvaged rails were used for fence posts, 1,324 linear feet for guardrails on the new roads, and 41,425 linear feet for a climb proof fence. (U.S. Army Photograph-Picatinny Arsenal.)

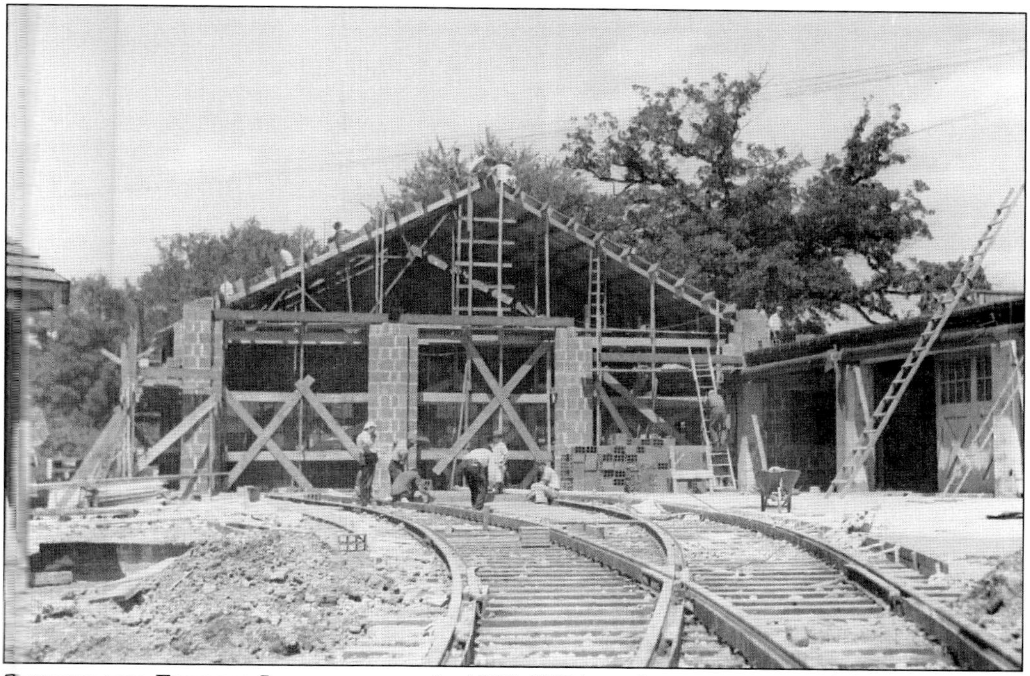

SHELTER FOR FIRELESS LOCOMOTIVES. In 1938, WPA workmen constructed this wooden and tile structure for the eight fireless locomotives used on the 2 miles of the arsenal's narrow gauge tracks. The locomotives, box and flat cars, and track were disposed of in the late 1970s. (U.S. Army Photograph-Picatinny Arsenal.)

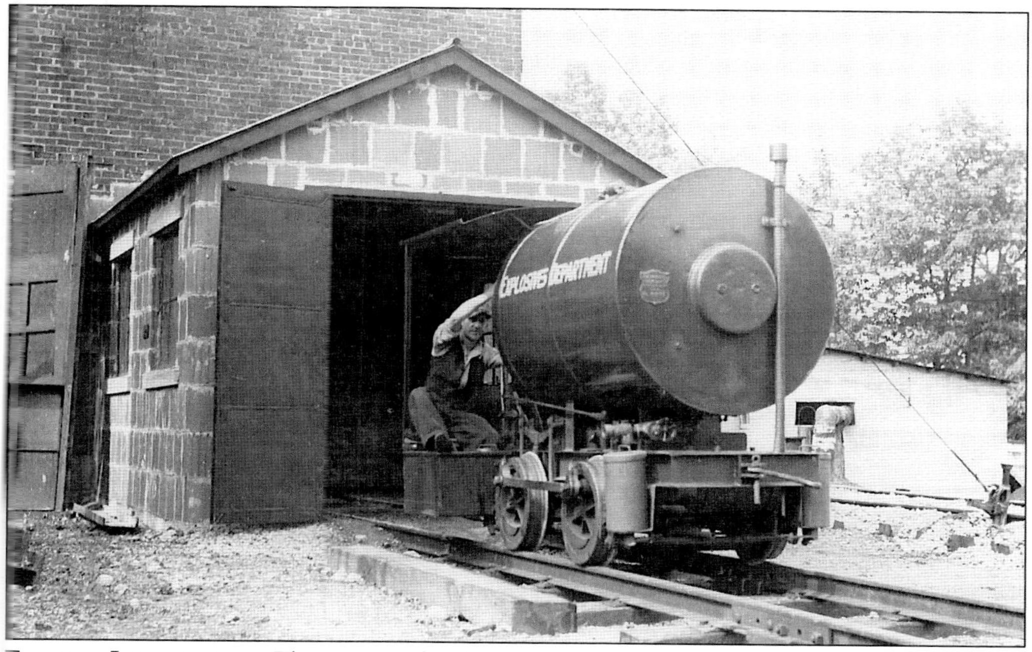

FIRELESS LOCOMOTIVE. This is one of eight fireless locomotives used to move munitions and explosives to the second floor of loading buildings. It then moved the finished munitions on the arsenal's narrow gauge track to waiting box cars and trucks for shipment to the fighting front. (U.S. Army Photograph-Picatinny Arsenal.)

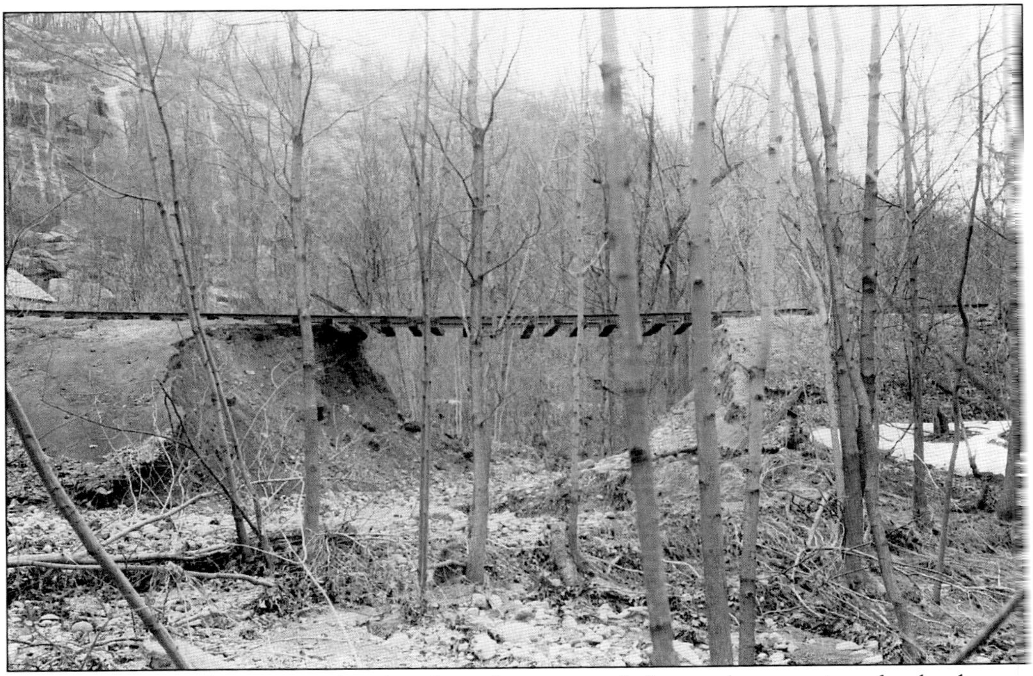

FLOOD DAMAGE. Shown are railroad tracks and ties suspended over a large section of embankment that was washed out in a 1936 flood that inundated much of the arsenal. The WPA constructed 41 culverts under both railroad and roads to prevent waters from Green Pond Brook from overflowing its banks and washing out tracks and roads. (U.S. Army Photograph-Picatinny Arsenal.)

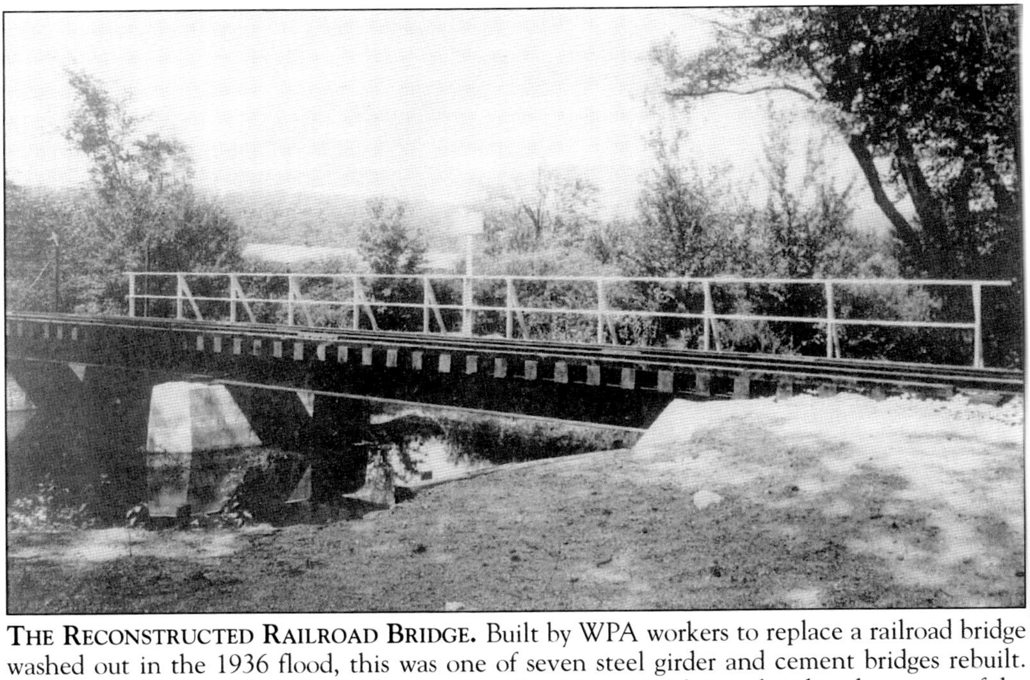

THE RECONSTRUCTED RAILROAD BRIDGE. Built by WPA workers to replace a railroad bridge washed out in the 1936 flood, this was one of seven steel girder and cement bridges rebuilt. The tracks rest on steel girders laid on cement abutments at either end and in the center of the bridge. (U.S. Army Photograph-Picatinny Arsenal.)

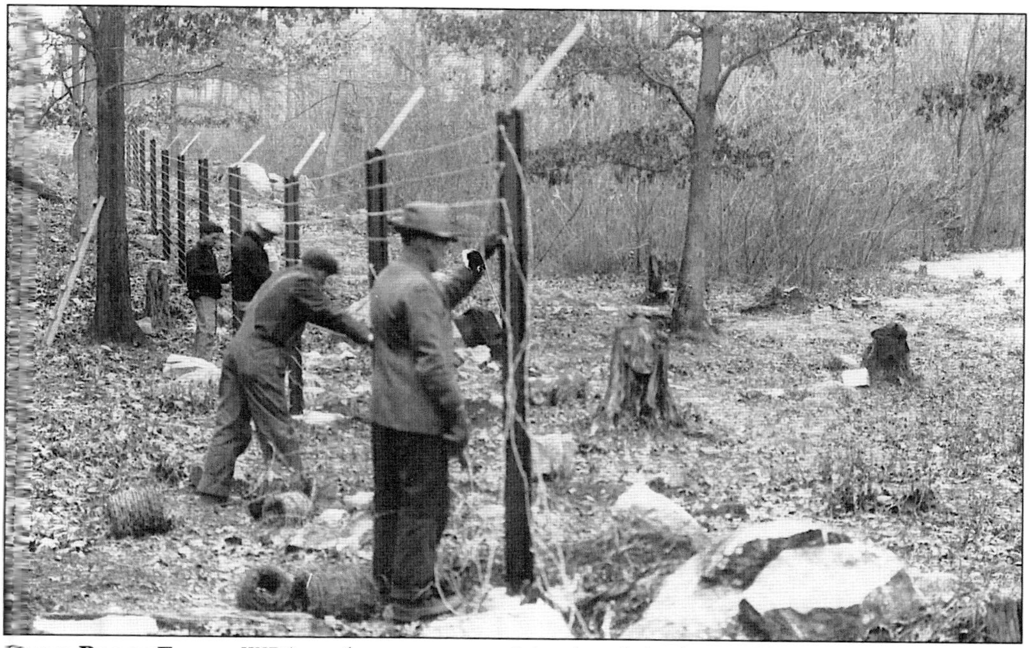

CLIMB PROOF FENCE. WPA workers constructed 6 miles of climb-proof fence around the arsenal to ward off trespassers. Workmen used iron rails salvaged from the 22,850 linear feet of outdated railroad tracks that were torn up and replaced. The posts were set in concrete, often on steep rock strewn hillsides. (U.S. Army Photograph-Picatinny Arsenal.)

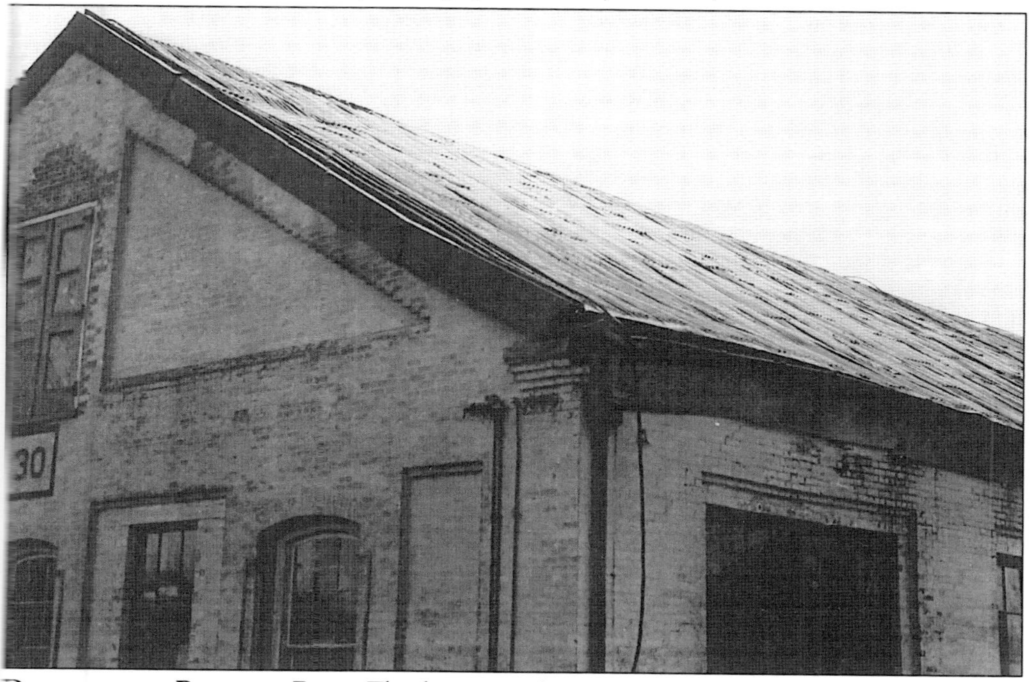

DETERIORATED BUILDING ROOF. The deteriorated condition of the roof on this brick storehouse is typical of the roofs replaced by WPA workmen on many storehouses and magazines. Built in 1890 at a cost of $24,847, this roof was part of 28,600 square feet of roofs replaced. (U.S. Army Photograph-Picatinny Arsenal.)

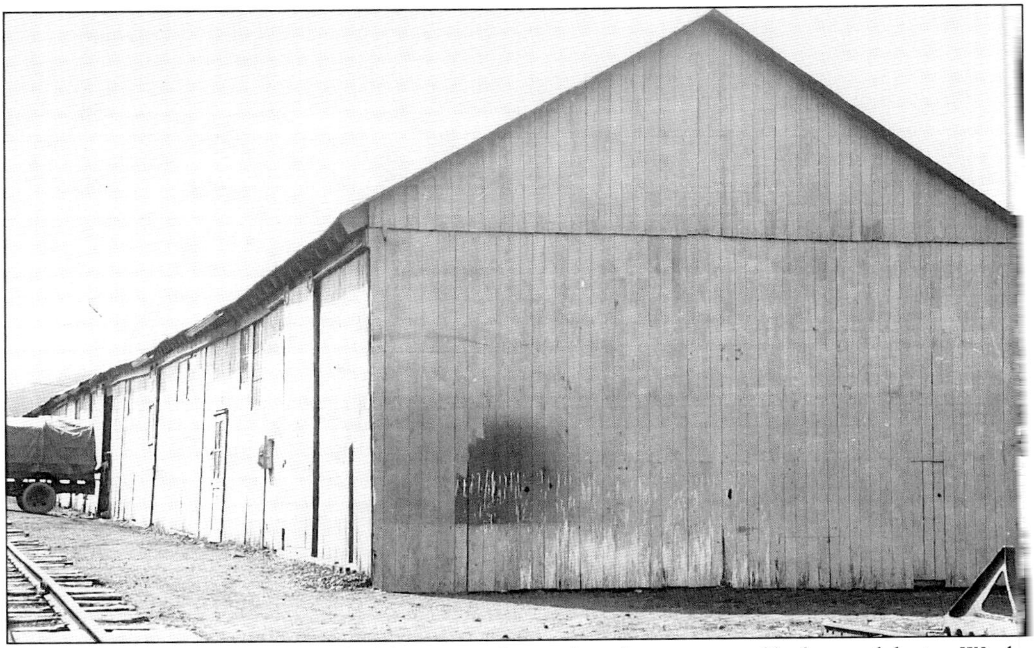

WOODEN STORAGE SHED. This building, one of many hastily constructed before and during World War I, is typical of storage facilities in use before WPA rehabilitation. Reconstructed were two 8-inch-tile storage buildings (27,000 square feet); four tile and lumber projectile storage buildings (90,000 square feet), and 102 smaller buildings. (U.S. Army Photograph-Picatinny Arsenal.)

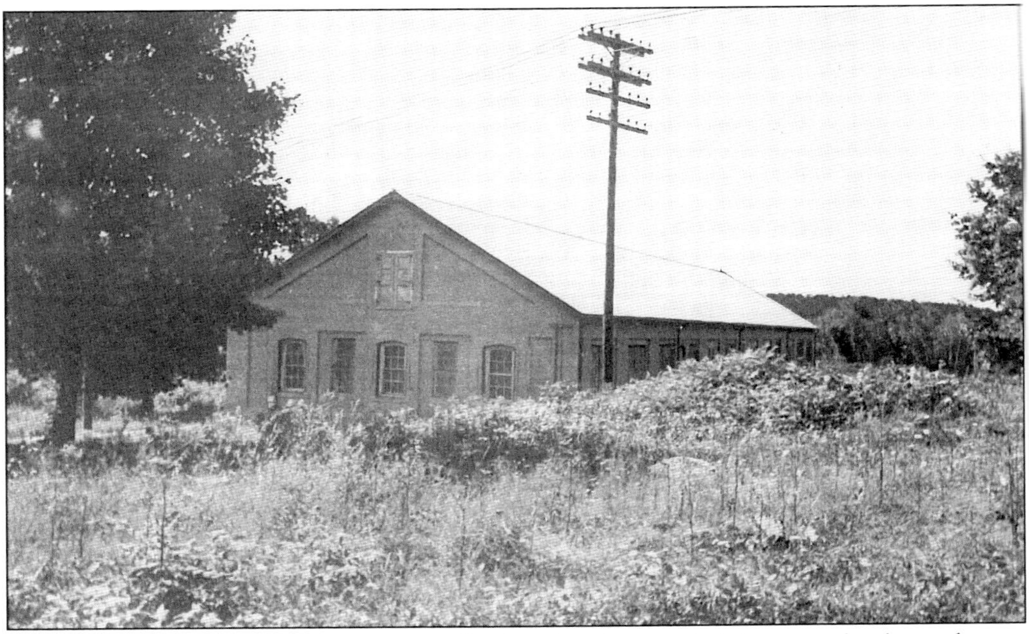

REHABILITATED BRICK STORAGE SHED. This building is typical of the seven brick warehouses (32,500 square feet) reconstructed and re-roofed by WPA workmen. It was one of the warehouses originally built at the turn of the century. (U.S. Army Photograph-Picatinny Arsenal.)

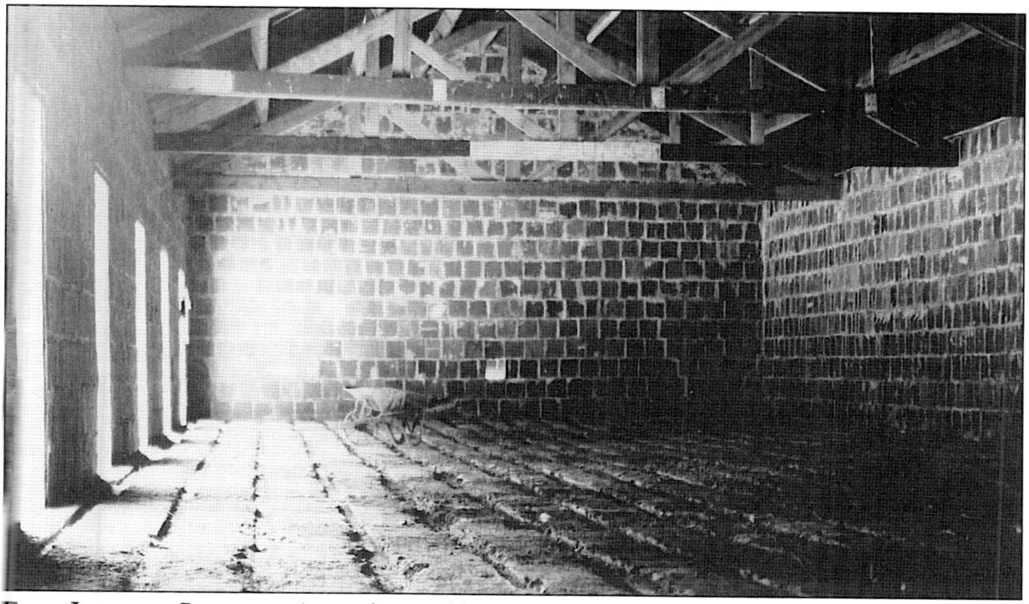

FACE LIFTING. Picatinny Arsenal was able to assume its role in World War II because of a $2.3 million congressional appropriation to rebuild buildings that were damaged in the 1926 explosion, and a $2.9 million face lift by 1,696 WPA workmen between 1932 and 1938. Shown here is the interior of a magazine, its rotting wooden floors removed, and the area cleared for installation of cement floors. (U.S. Army Photograph-Picatinny Arsenal.)

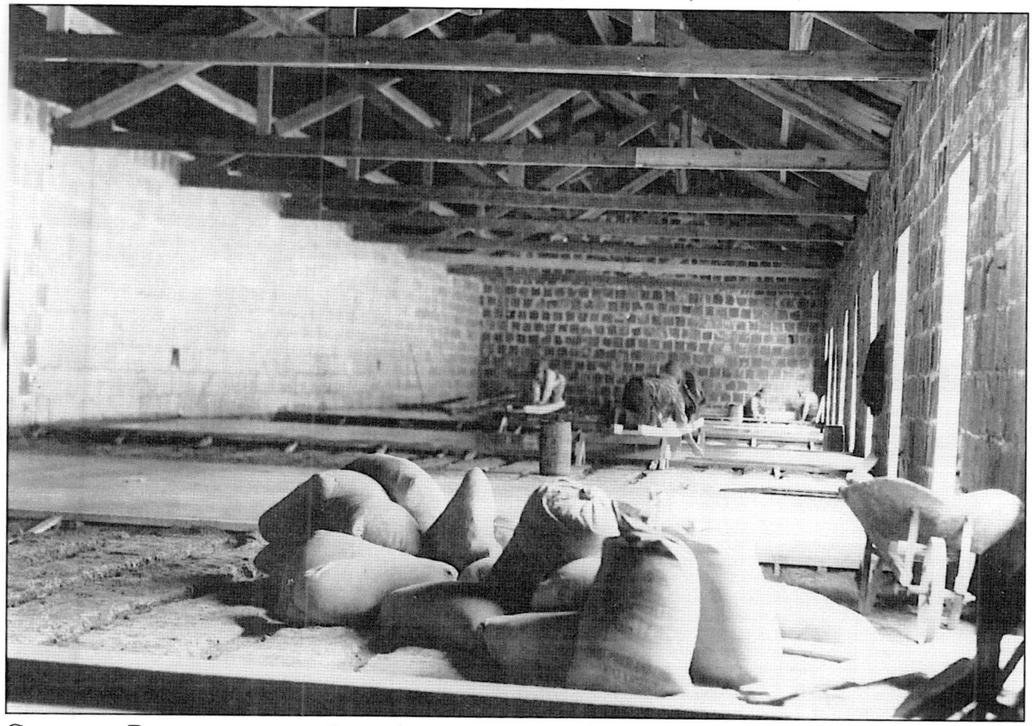

COMPLETE REHABILITATION. Sixty-five magazines with 8-inch-thick-tile walls (157,000 square feet) were completely rehabilitated by the WPA work force, in many cases it included new cement floors and roofs. (U.S. Army Photograph-Picatinny Arsenal.)

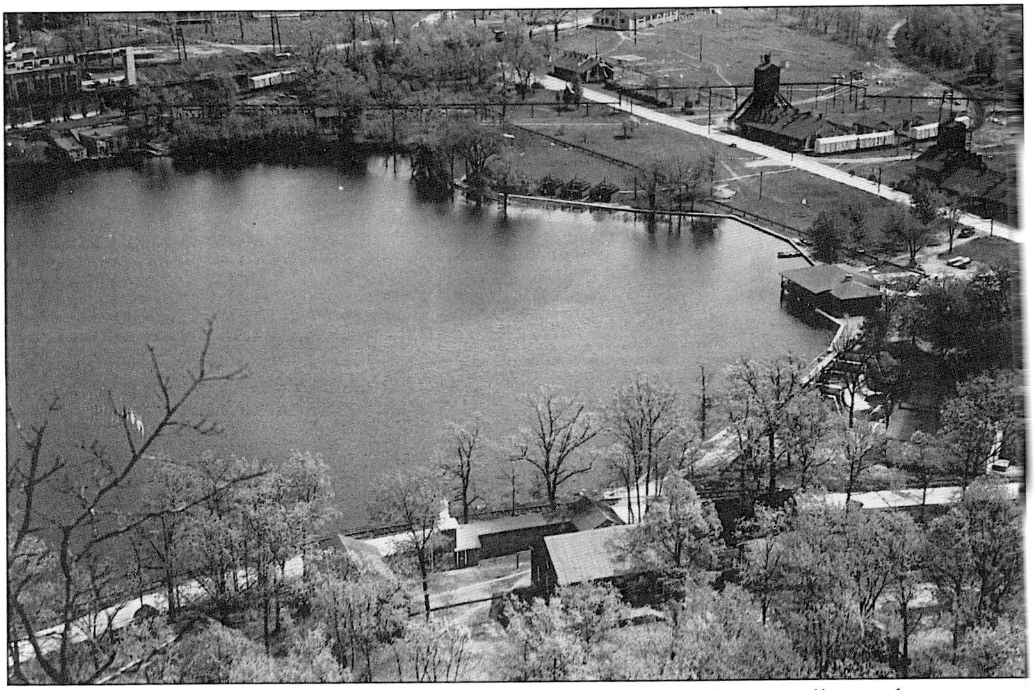

Aerial View of the New Seawall. This aerial view shows the seawall stretching out on both sides of the dam. The first dam, a crude small affair, was built in 1749 by Jacob Osborne to provide power for his forge. Later, after the army established the arsenal in 1880, a waterwheel on a larger dam provided power. (U.S. Army Photograph-Picatinny Arsenal.)

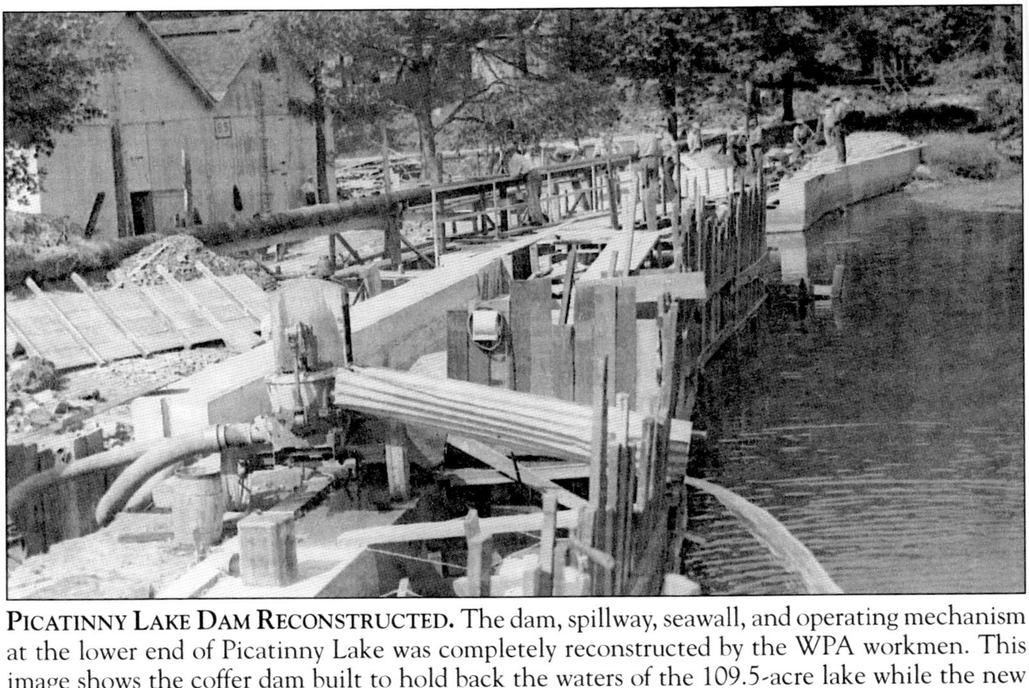

Picatinny Lake Dam Reconstructed. The dam, spillway, seawall, and operating mechanism at the lower end of Picatinny Lake was completely reconstructed by the WPA workmen. This image shows the coffer dam built to hold back the waters of the 109.5-acre lake while the new dam and spillway was under construction. (U.S. Army Photograph-Picatinny Arsenal.)

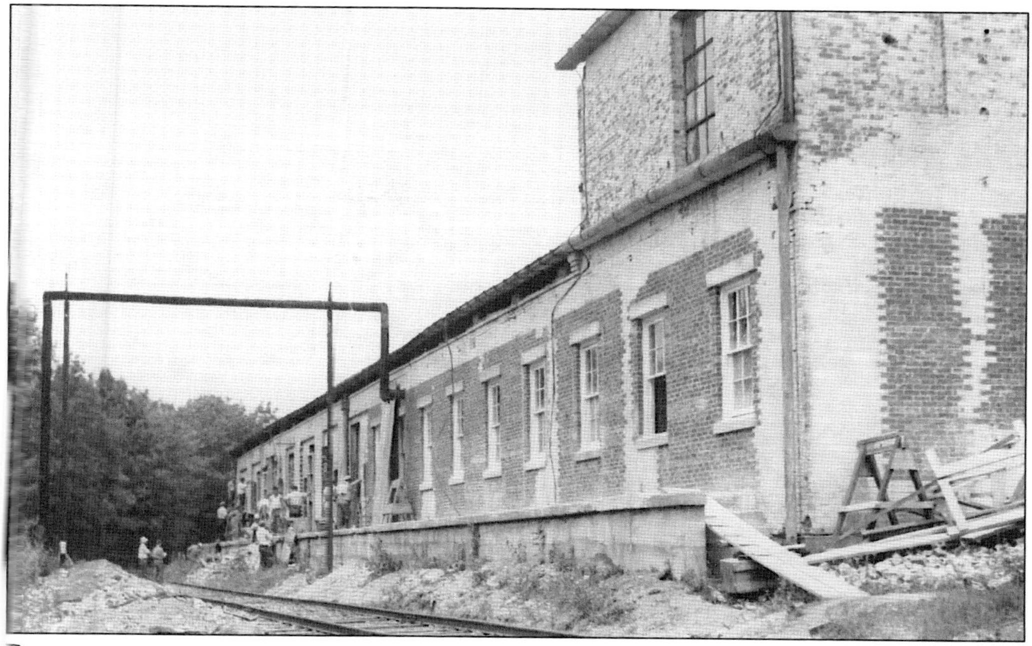

RECONSTRUCTION OF THE FIRST FLOOR OF THE EXPERIMENTAL FUSE PLANT. WPA masons reconstructed brickwork on the first floor of a brick experimental fuse plant for the development of fuses used to detonate bombs. It was at the arsenal that special fuses for the first bombing of Tokyo and the bombing of the Ploesti oil fields were developed and manufactured. (U.S. Army Photograph-Picatinny Arsenal.)

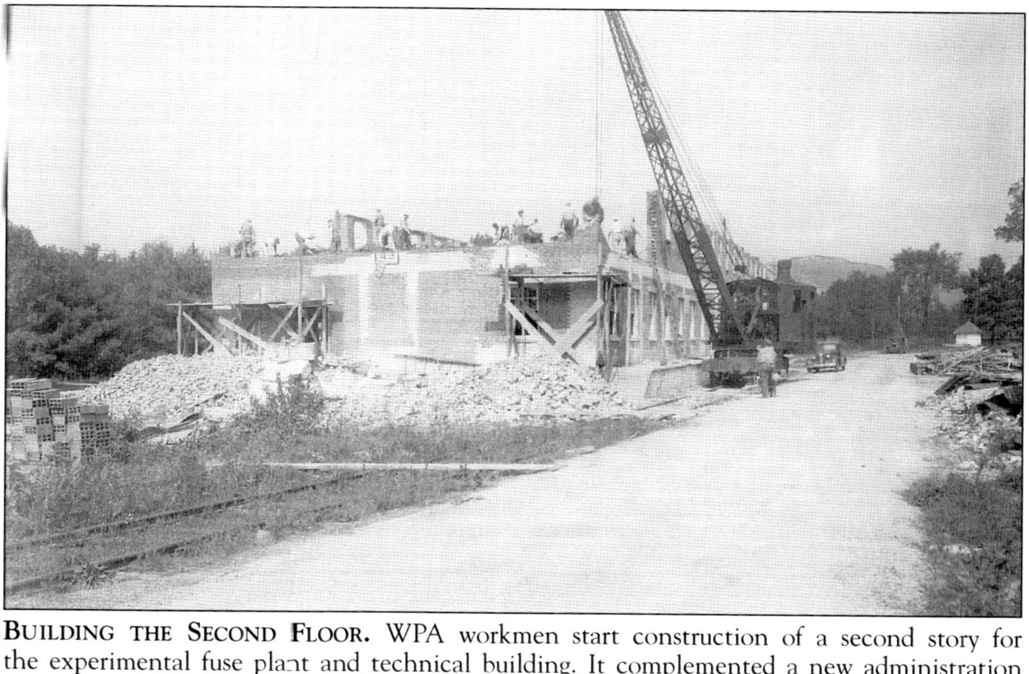

BUILDING THE SECOND FLOOR. WPA workmen start construction of a second story for the experimental fuse plant and technical building. It complemented a new administration center, laboratories, magazines, an 8,000-square-foot pyrotechnic factory and plating plant, sand blastings, shell cleaning building, and experimental powder factory. (U.S. Army Photograph-Picatinny Arsenal.)

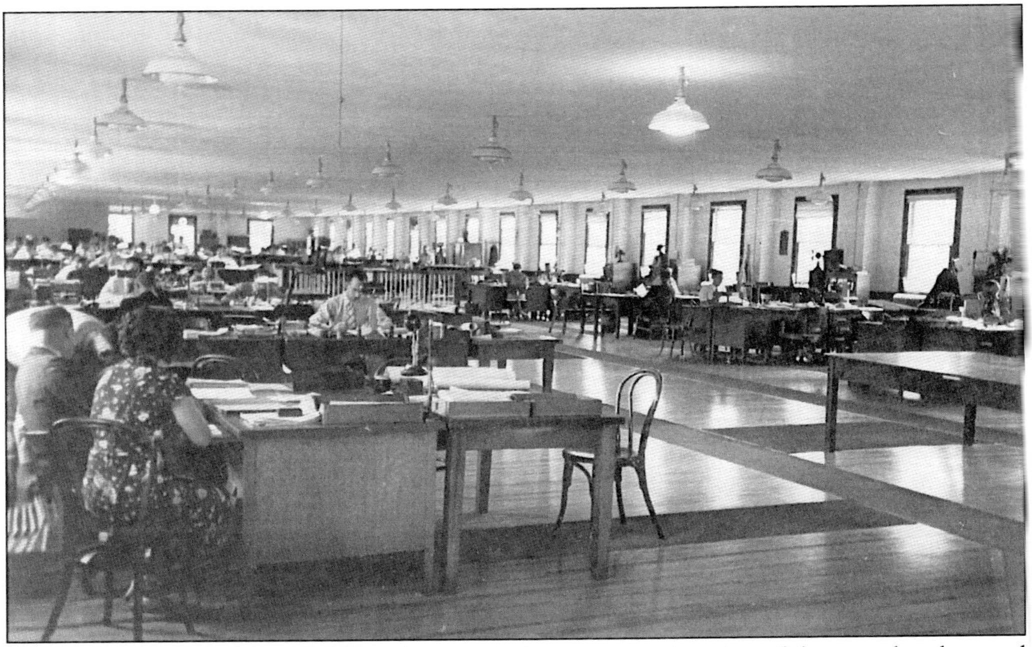

Completed Second Floor. This photograph shows an interior view of the completed second floor of the experimental fuse plant and technical building with the staff at work. Other major projects completed by the WPA included a modern brick officers quarters, an officer's club, employee's athletic club, a 4,300-square-foot greenhouse, and a rock crushing plant. (U.S. Army Photograph-Picatinny Arsenal.)

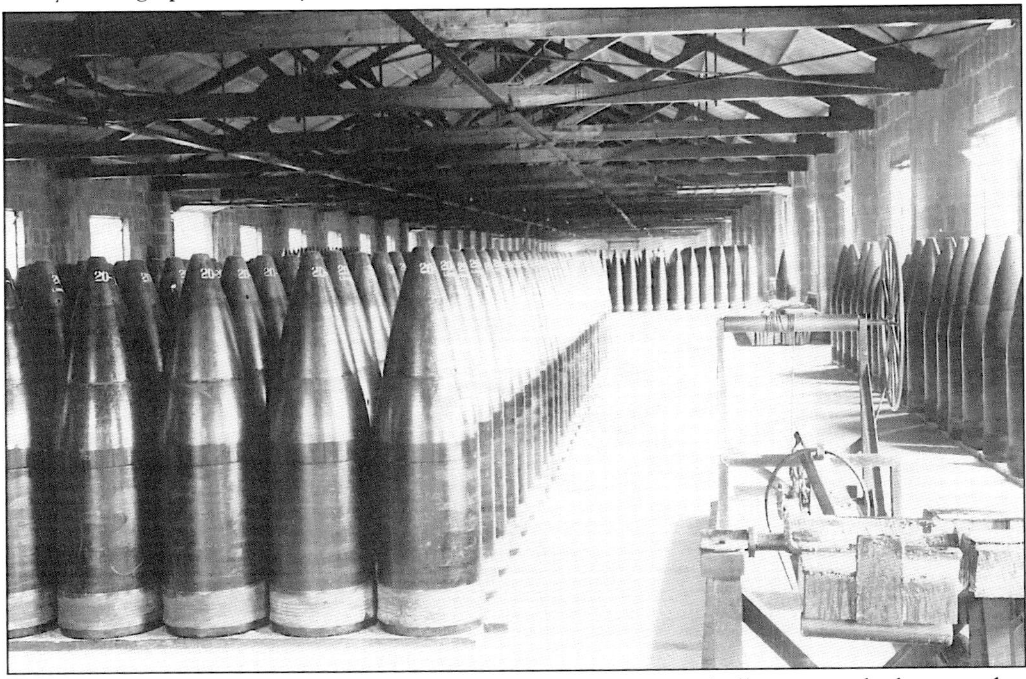

The Completed Storage Shed. Rows of large-caliber artillery shells were stacked on wooden pallets in a storage shed, rebuilt in the 1930s by WPA workmen after the 1926 explosion. (U.S. Army Photograph-Picatinny Arsenal.)

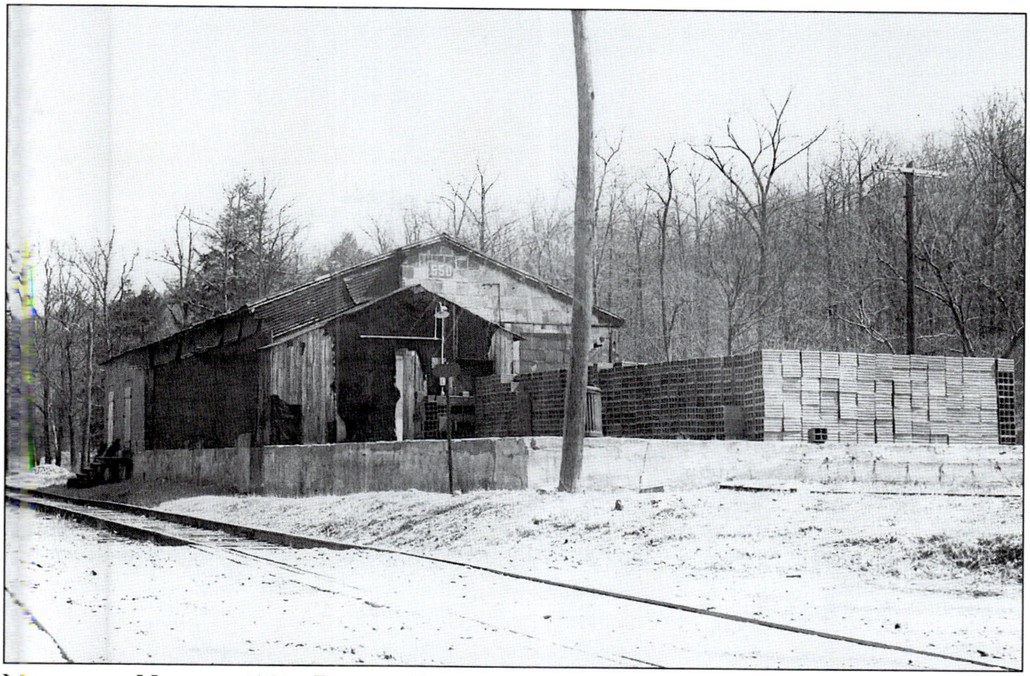

MAGAZINE NUMBER 550, BEFORE REHABILITATION. A tile magazine from *c.* 1900, this building was readied for rehabilitation by WPA workmen. Piles of 8-inch-thick tiles brought to the site by railroad cars were stacked on the cement foundation ready for masons to start the work that would double the size of the magazine. (U.S. Army Photograph-Picatinny Arsenal.)

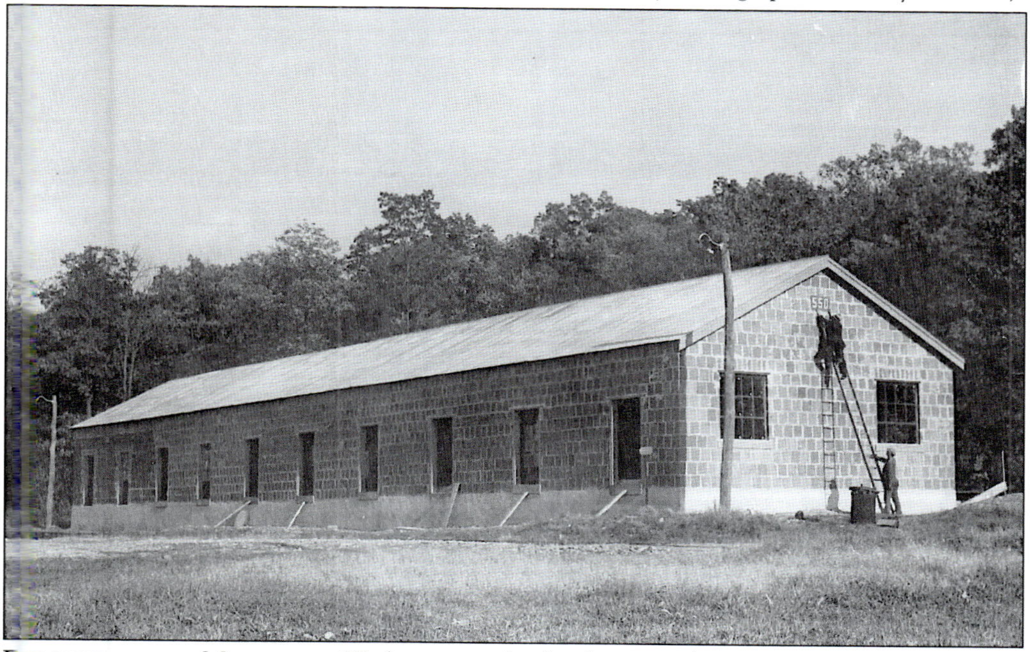

RECONSTRUCTED MAGAZINE. Workmen put the finishing touches on building number 550, a tile-walled magazine that suffered damage in the 1926 explosion. The wooden addition partially covered with tar paper and the corrugated metal over the blocked-off doors covered with tar paper were removed and a new roof installed. (U.S. Army Photograph-Picatinny Arsenal.)

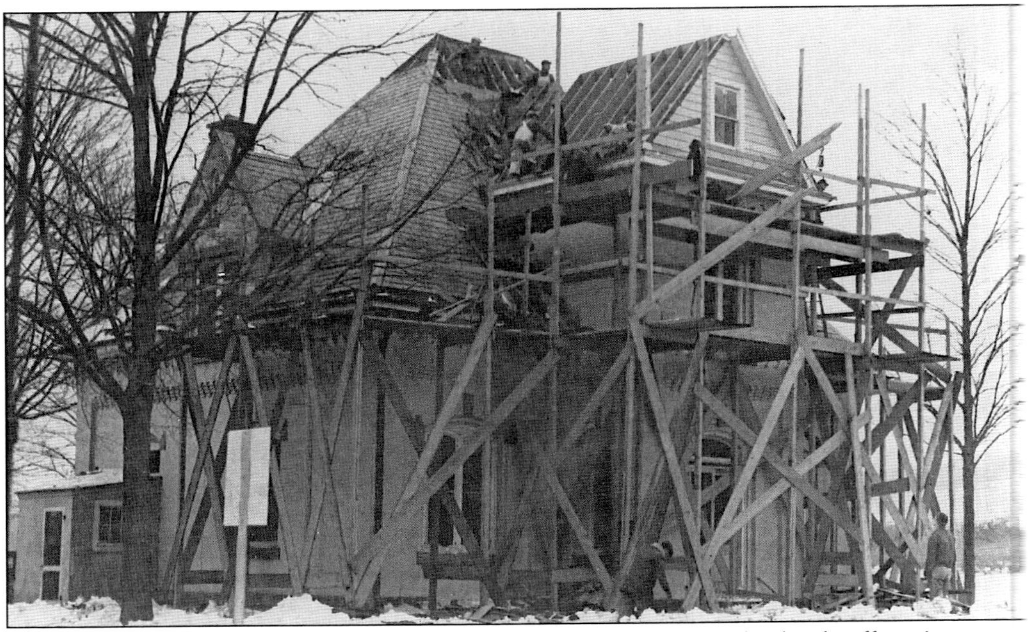

Officers' Quarters. WPA carpenters and masons reconstruct the brick officers' quarters that was badly damaged by shock waves and deterioration in the 1926 explosion. A typical Victorian-era residence, similar in architecture to the chief of staff's residence, its appearance was completely changed by the facelift. (U.S. Army Photograph-Picatinny Arsenal.)

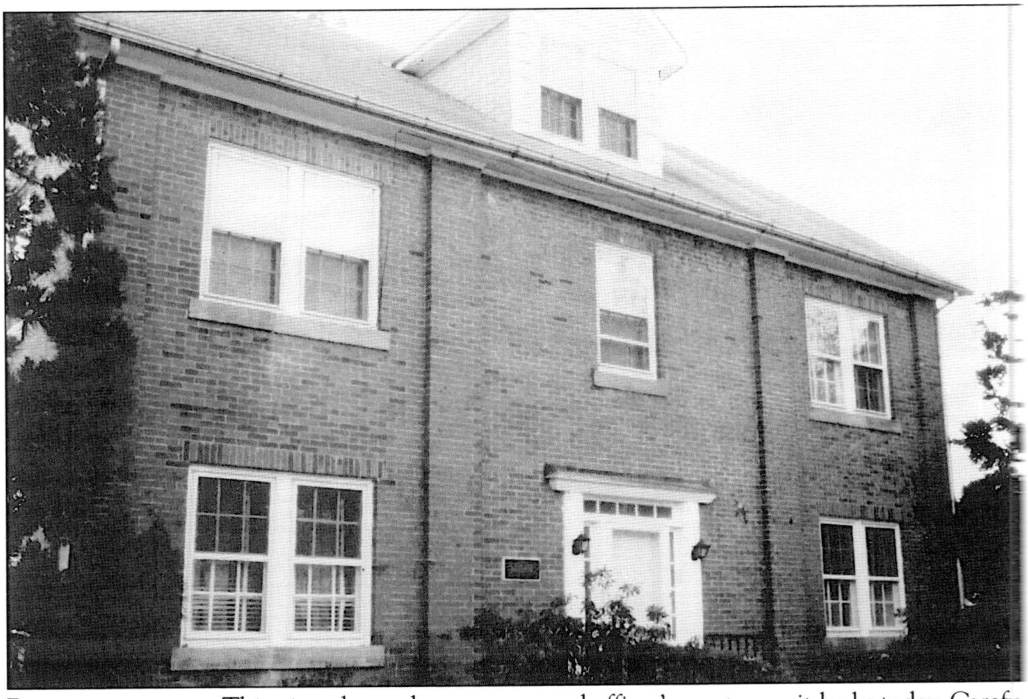

Reconstruction. This view shows the reconstructed officer's quarters as it looks today. Careful examination reveals brick of two shades: the original red brick predating World War I salvaged by WPA masons and a brighter colored new brick. (U.S. Army Photograph-Picatinny Arsenal.)

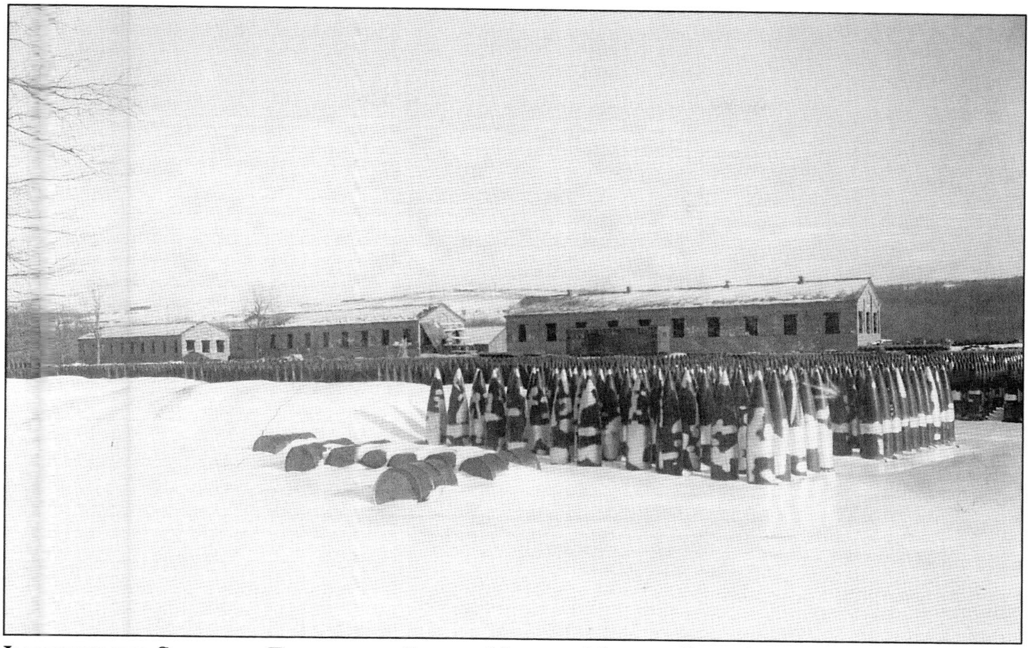

INADEQUATE STORAGE FACILITIES. Rows of large-caliber artillery shells were stacked in the open as far as the eye could see for lack of interior storage space, much of which was destroyed by the 1926 explosion. It was several years before the damaged warehouses and magazines were replaced. (U.S. Army Photograph-Picatinny Arsenal.)

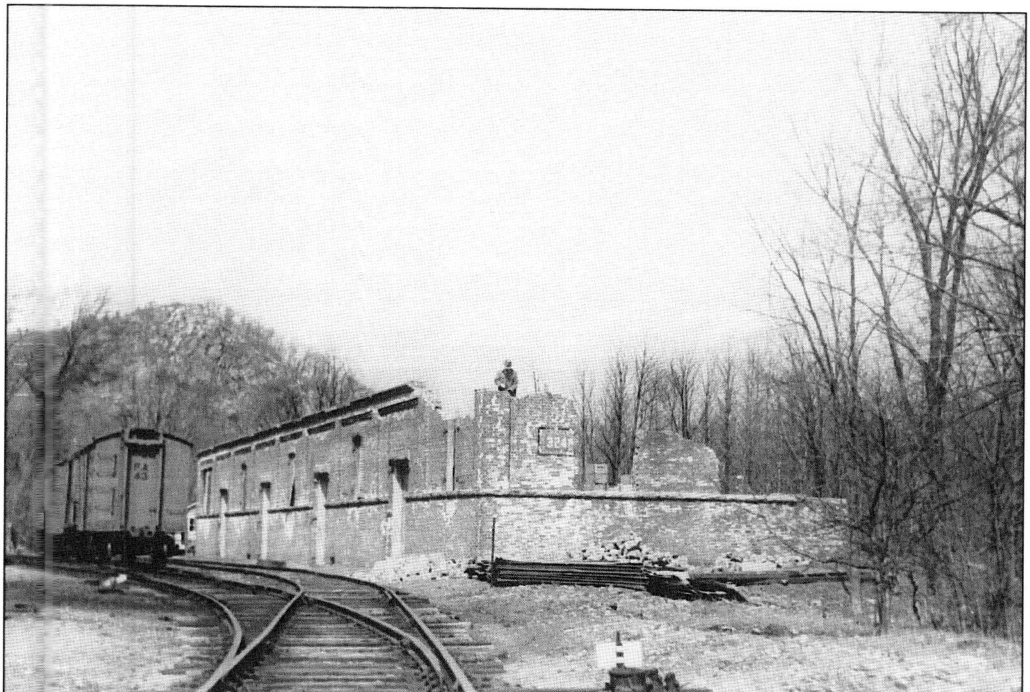

RECONSTRUCTION. WPA workmen started reconstructing building 324. The brick walls on one side and both ends collapsed from the force of the 1926 explosion, causing the roof trusses to collapse and the roof to cave in. (U.S. Army Photograph-Picatinny Arsenal.)

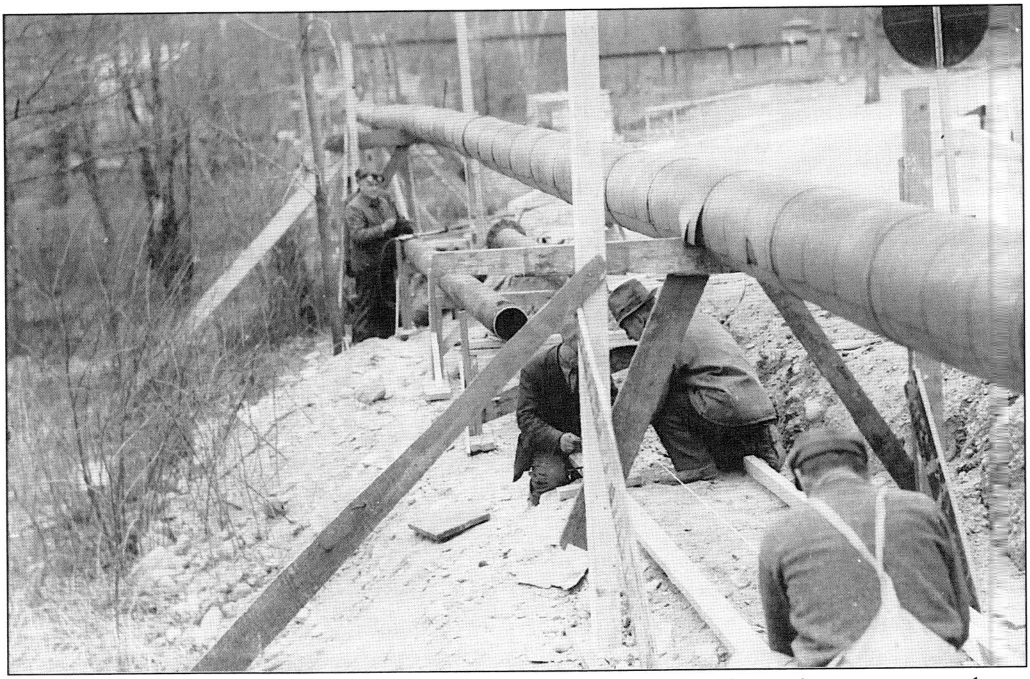

STEAM LINES. WPA workmen relocated and built new steam lines that criss-crossed the arsenal, bringing heat to buildings from the powerhouse. A total of 3,650 linear feet of outside steam lines using 1,802 steel posts and banjo hangers for support were rehabilitated. (U.S. Army Photograph-Picatinny Arsenal.)

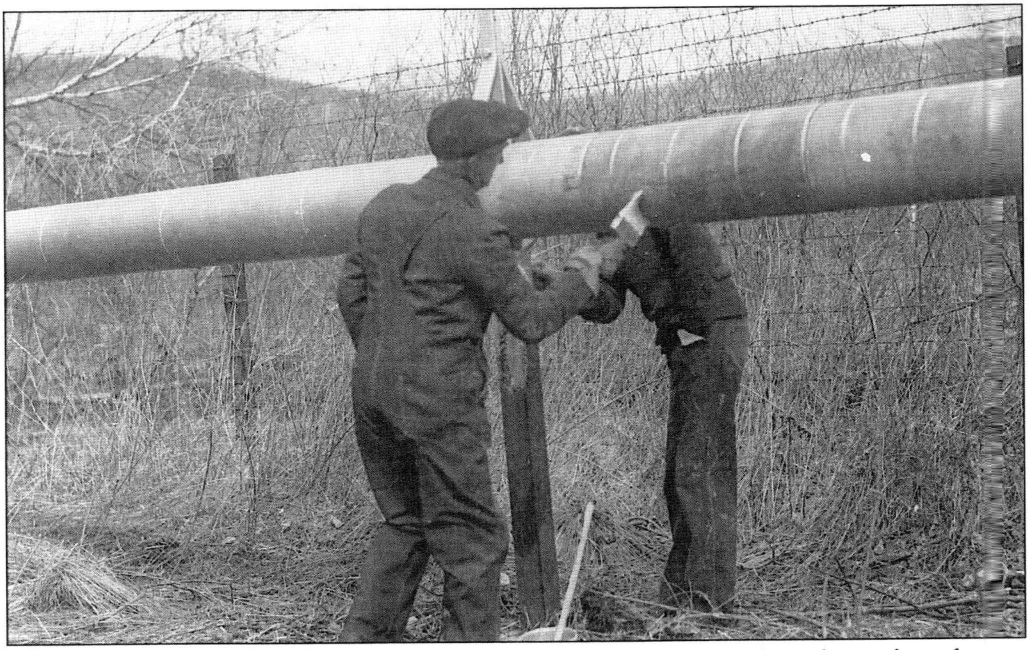

WEATHERPROOF STEAM LINES. WPA workmen covered more than 15,000 linear feet of steam lines with a weatherproof jacket and paint. Some of the rehabilitated steam lines were at ground level; others entered buildings two stories in the air. (U.S. Army Photograph-Picatinny Arsenal.)

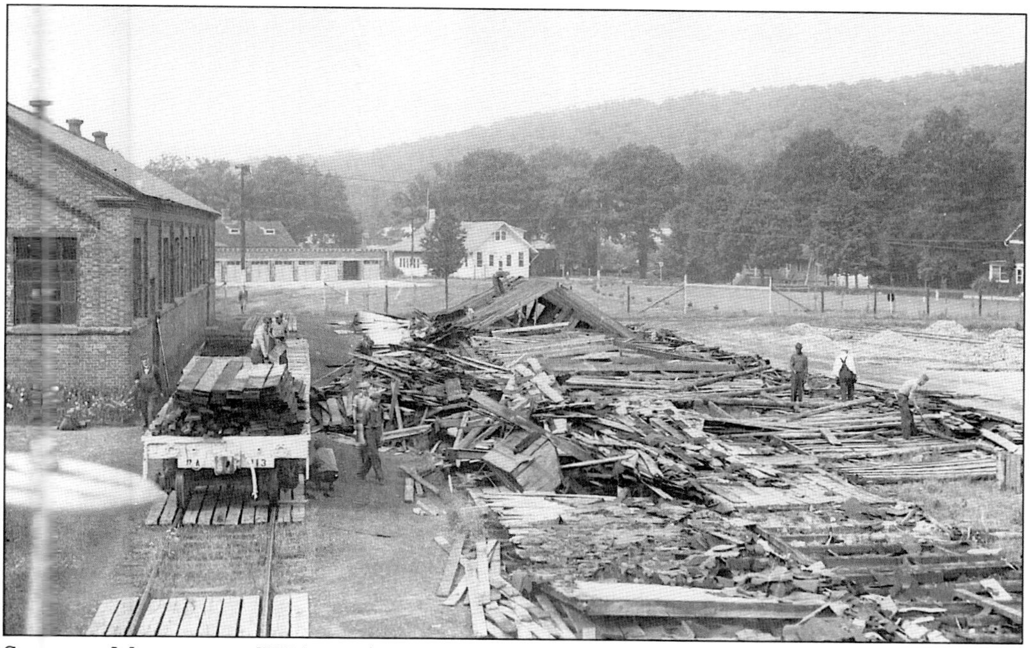

SALVAGE MATERIALS. WPA workmen salvaged materials from a frame warehouse damaged in the 1926 explosion. Boards, many already stacked on a railroad flatcar, were used in the rehabilitation of other buildings. (U.S. Army Photograph-Picatinny Arsenal.)

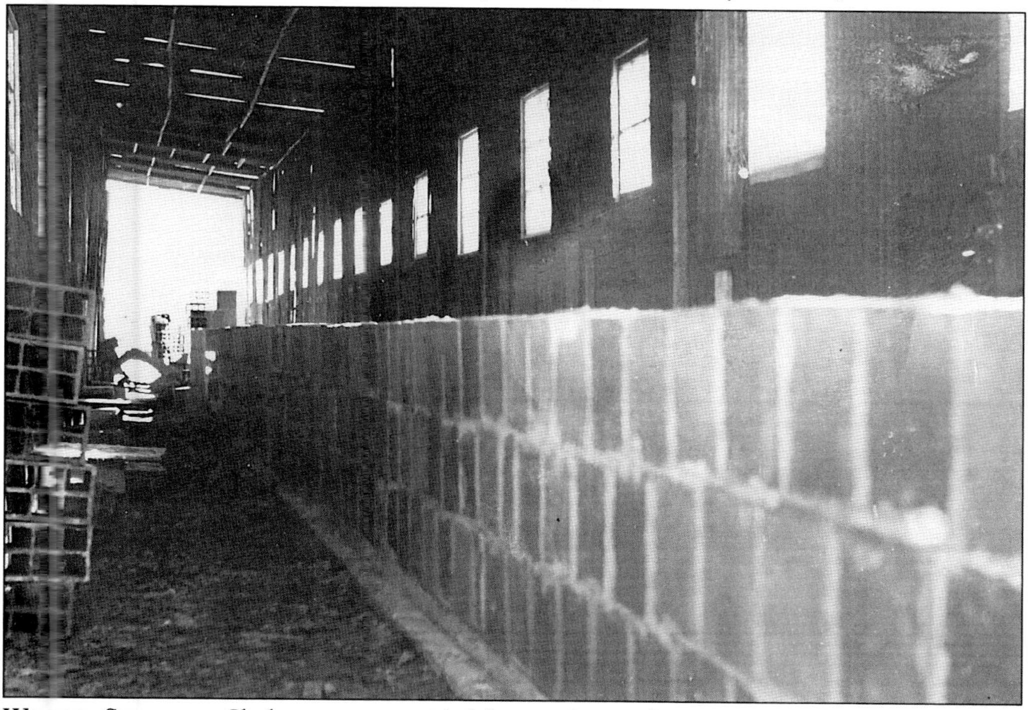

WINTER SHELTERS. Shelters were provided for use during the winter months so that work by WPA workmen would not be halted by snow, ice, and cold. Shown here is an 8-inch-thick-tile wall under construction during the winter. The work done by the WPA was valued at $2,769,800. (U.S. Army Photograph-Picatinny Arsenal.)

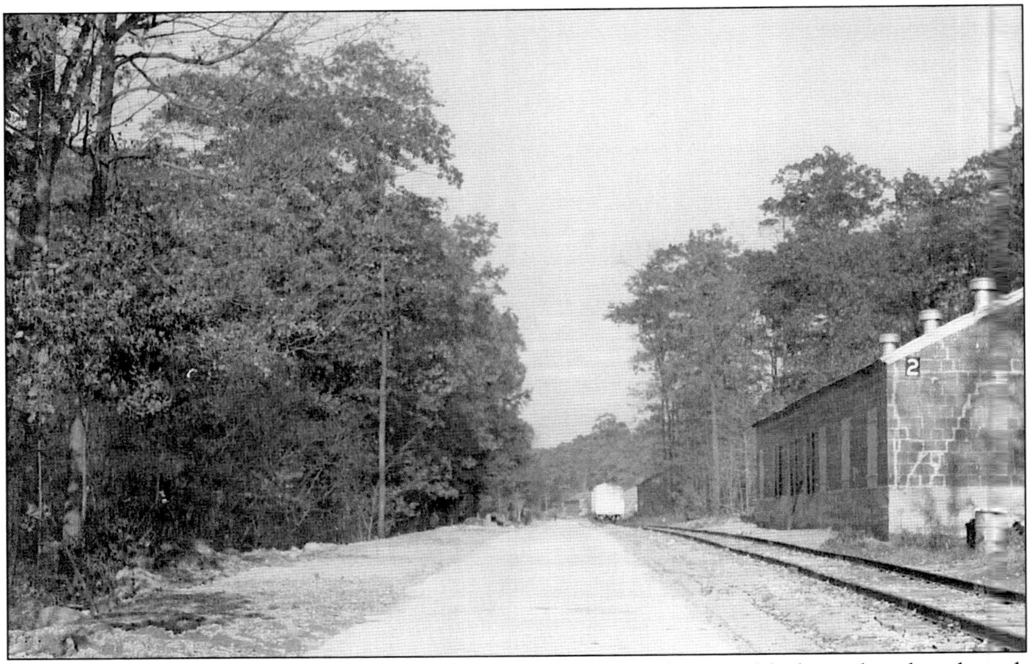

CRACKS IN TILE MAGAZINE WALLS. Screened by trees and accessible by railroad and road, warehouse number 2 shows evidence on its end walls of cracks sustained from the 1926 explosion. To prevent future damage to critical magazines, the WPA constructed eight barricades in the high explosive magazine area. (U.S. Army Photograph-Picatinny Arsenal.)

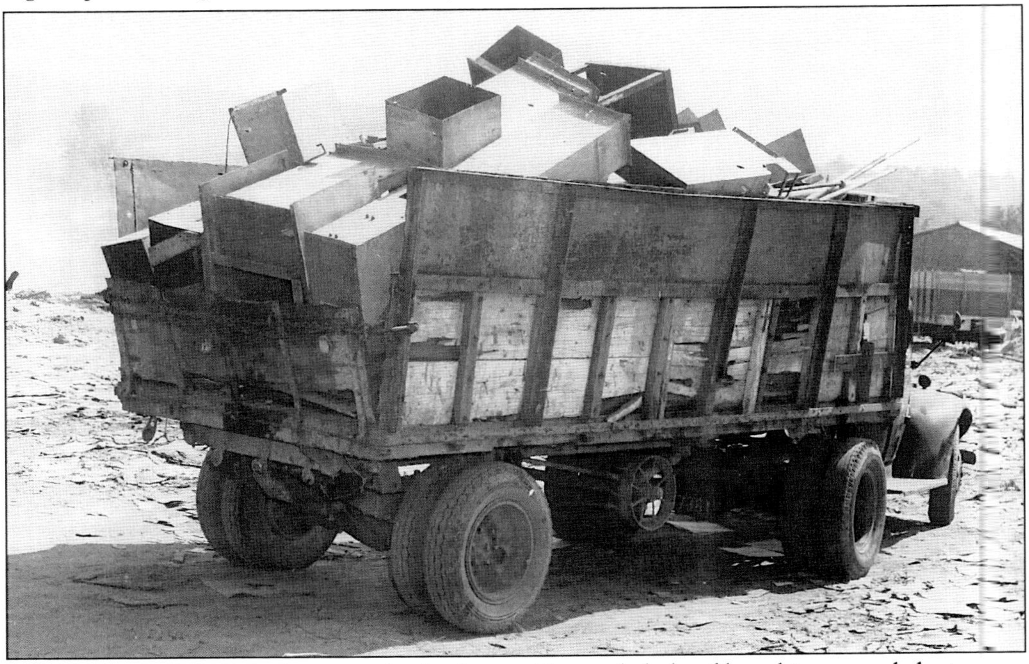

THE JUNKMAN CALLS. A wagon load of unusable steel desks, file cabinets, and ducts were carted away by a junkman. The arsenal sold all of its obsolescent and badly damaged items to junkmen, using funds received to purchase new items and rehabilitate buildings. (U.S. Army Photograph-Picatinny Arsenal.)

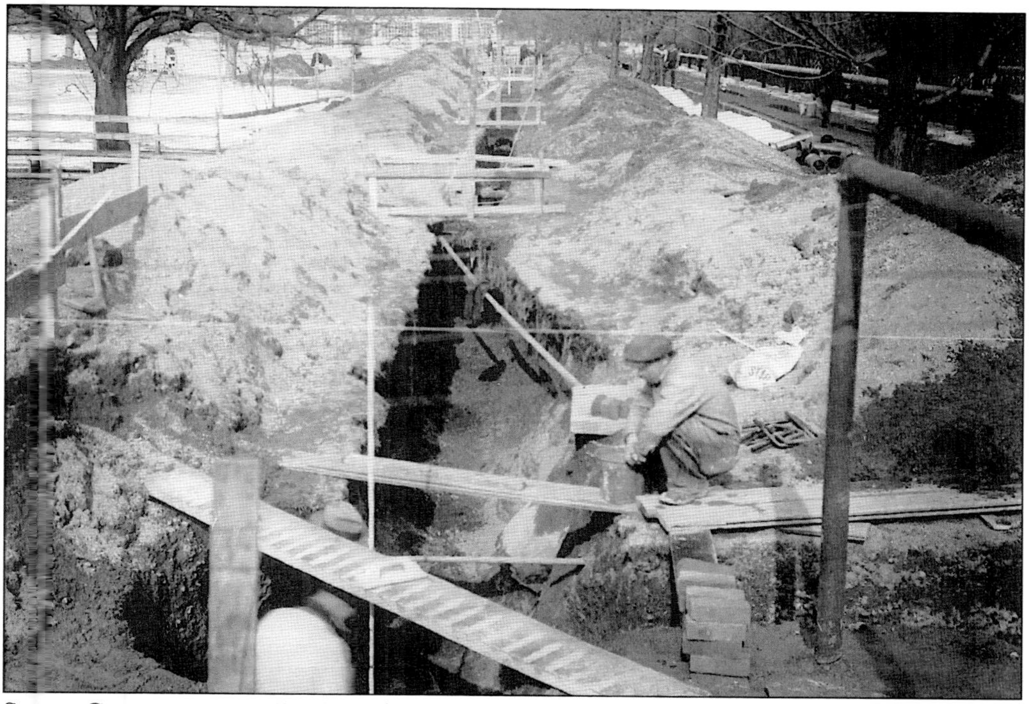

SEWER CONSTRUCTION. WPA workmen installed 2,700 linear feet of new sewer lines in deep trenches at the arsenal. In addition, they rehabilitated an outdated sewage treatment plant, activities that materially improved sewage facilities. (U.S. Army Photograph-Picatinny Arsenal.)

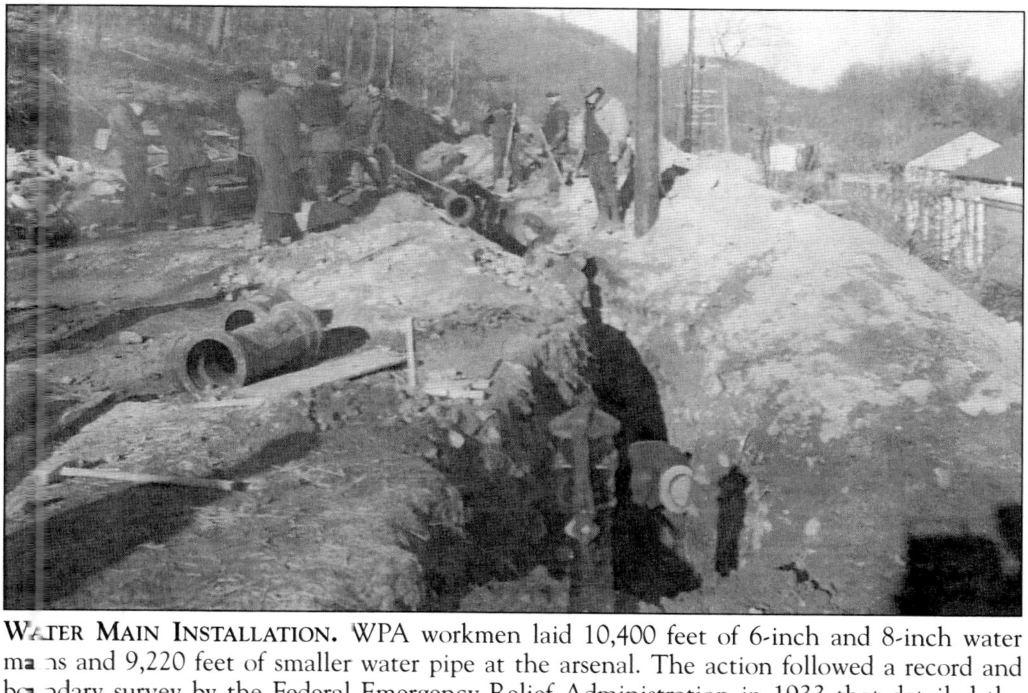

WATER MAIN INSTALLATION. WPA workmen laid 10,400 feet of 6-inch and 8-inch water mains and 9,220 feet of smaller water pipe at the arsenal. The action followed a record and boundary survey by the Federal Emergency Relief Administration in 1933 that detailed the needs of the arsenal. (U.S. Army Photograph-Picatinny Arsenal.)

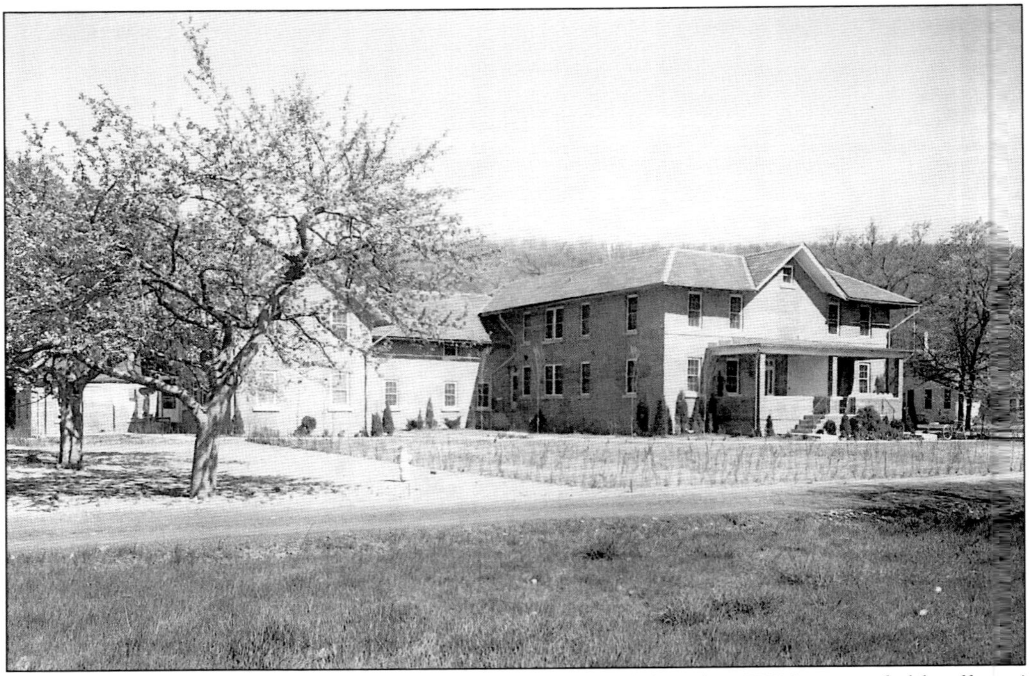

BUILDING CONVERSION. This building was converted by the WPA into field officers' apartments, transients' quarters, and a recreation room. Work started in 1933 when the Civilian Works Administration sent 417 men to survey roads and bridges, build new tar roads, improve old macadam roads, dredge Green Pond Brook, salvage buildings, and build a dam across Bear Creek Swamp Brook. (U.S. Army Photograph-Picatinny Arsenal.)

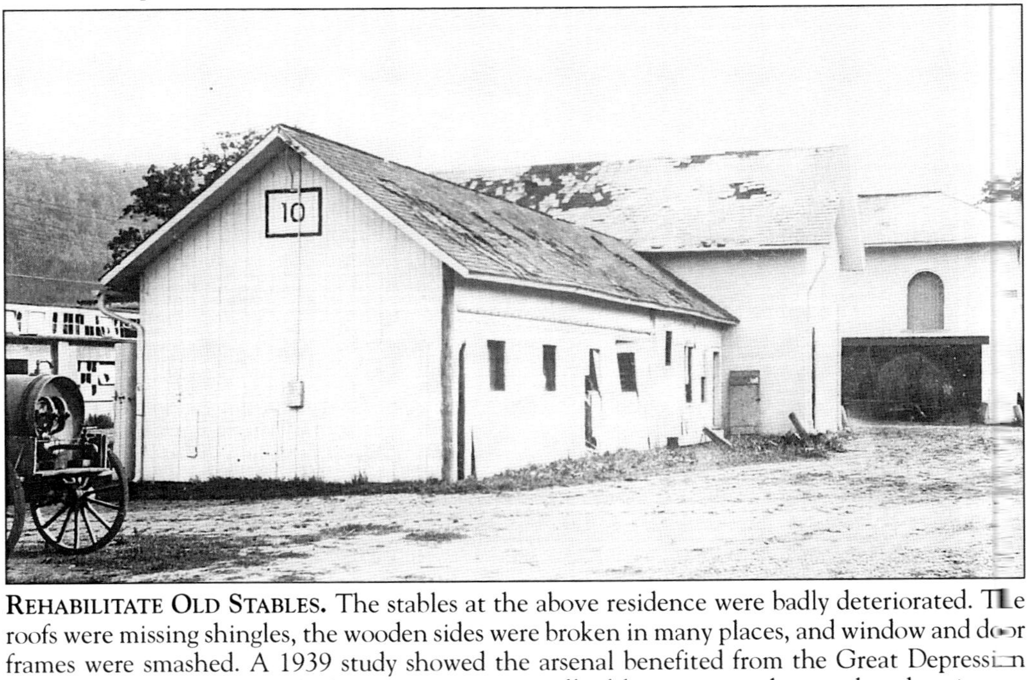

REHABILITATE OLD STABLES. The stables at the above residence were badly deteriorated. The roofs were missing shingles, the wooden sides were broken in many places, and window and door frames were smashed. A 1939 study showed the arsenal benefited from the Great Depression because relief agencies supplied, at no cost, services of highly experienced unemployed engineers. (U.S. Army Photograph-Picatinny Arsenal.)

Four

WORLD WAR II

PICATINNY SUPPLIES THE WORLD

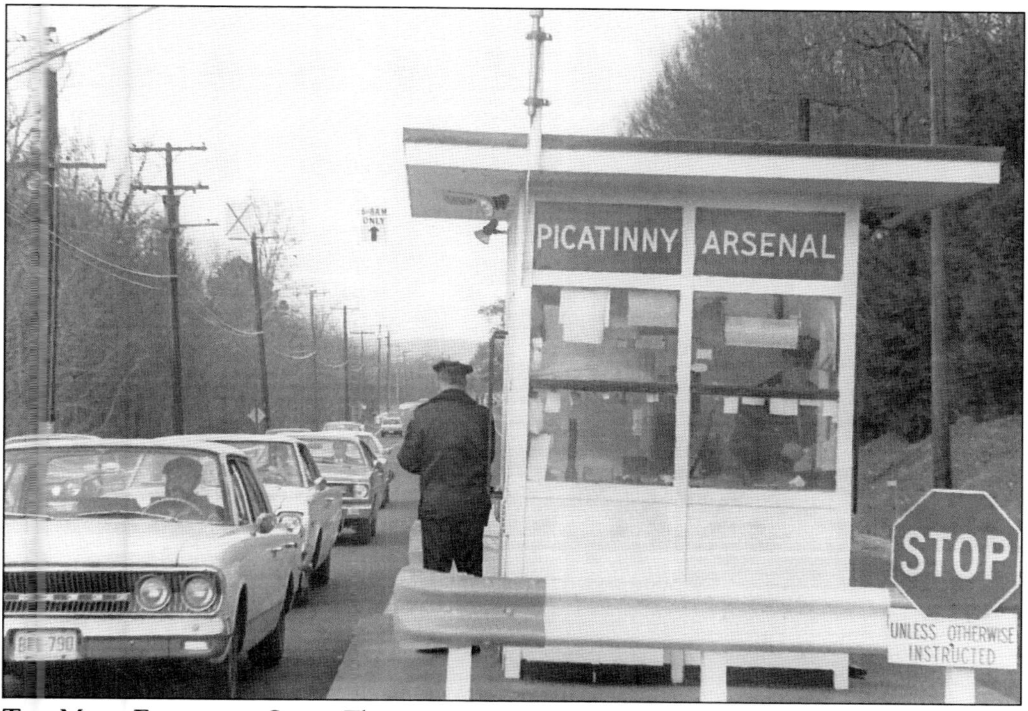

THE MAIN ENTRANCE GATE. The main gate to Picatinny Arsenal, off Rt. 15, is where cars and occupants were checked by civilian guards when entering and leaving the arsenal. It led to the three-lane Parker Road, named for the initial commanding officer at Picatinny, and the Cannon Gates, 1 mile distant. (U.S. Army Photograph-Picatinny Arsenal.)

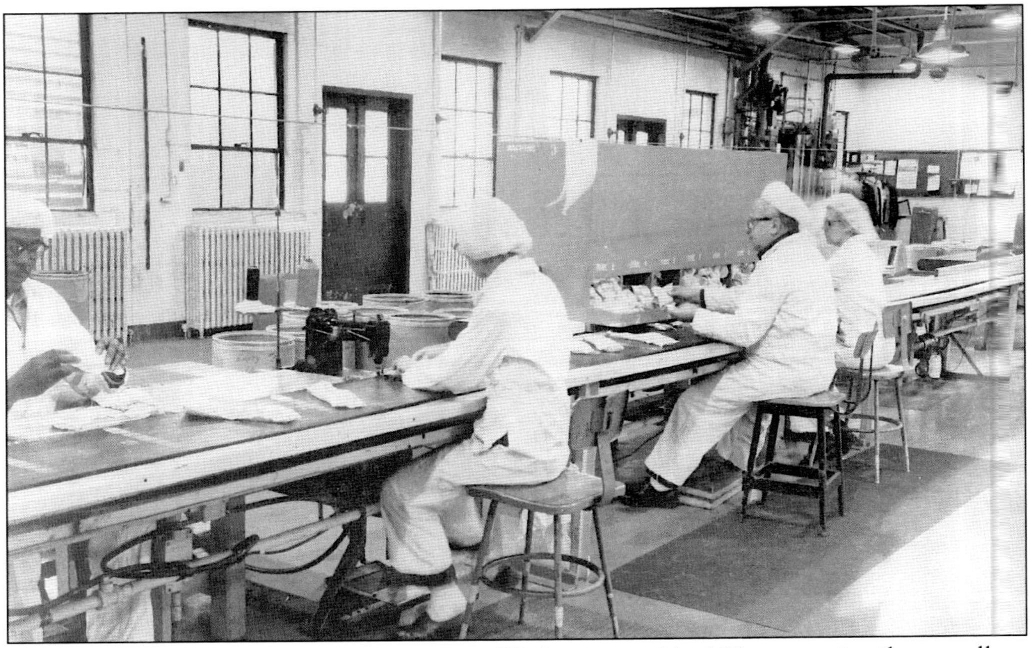

World War II Powder Bag Assembly. Workers assemble 155-mm projectile propellant bags from cotton material developed at the arsenal to replace silk bags after that material became hard to obtain. Especially designed to handle large quantities of propellant powder, the assembly area included 20 buildings comprising five major assembly units. (U.S. Army Photograph-Picatinny Arsenal.)

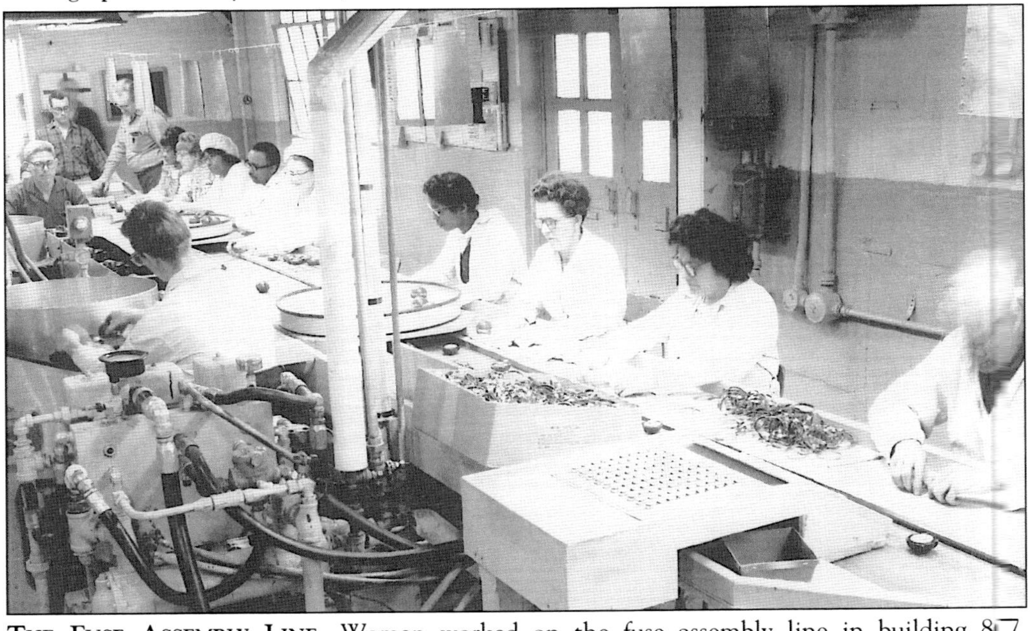

The Fuse Assembly Line. Women worked on the fuse assembly line in building 8[illegible]7, constructed in 1930 as part of a line of buildings 2,400 feet long with connecting ramps. The line loaded, assembled, and packed complete rounds for shipment, in addition to shells for semi-fixed rounds, separate loaded shells, fragmentation, and demolition bombs. (U.S. Army Photograph-Picatinny Arsenal.)

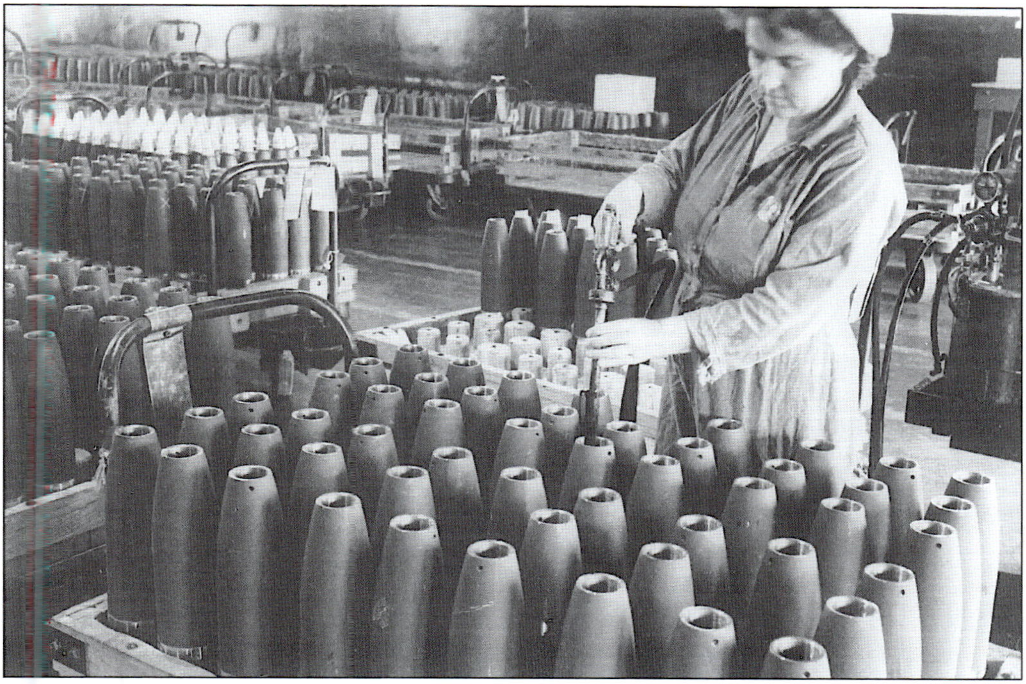

Loading 155-mm Artillery Shells. A woman loads 155-mm artillery shells with liquid explosive, which hardens in the shell case. Platforms containing 54 shell casing each fill the loading room. Each shell, before shipment to the fighting front, received two coats of paint plus identifying markings. (U.S. Army Photograph-Picatinny Arsenal.)

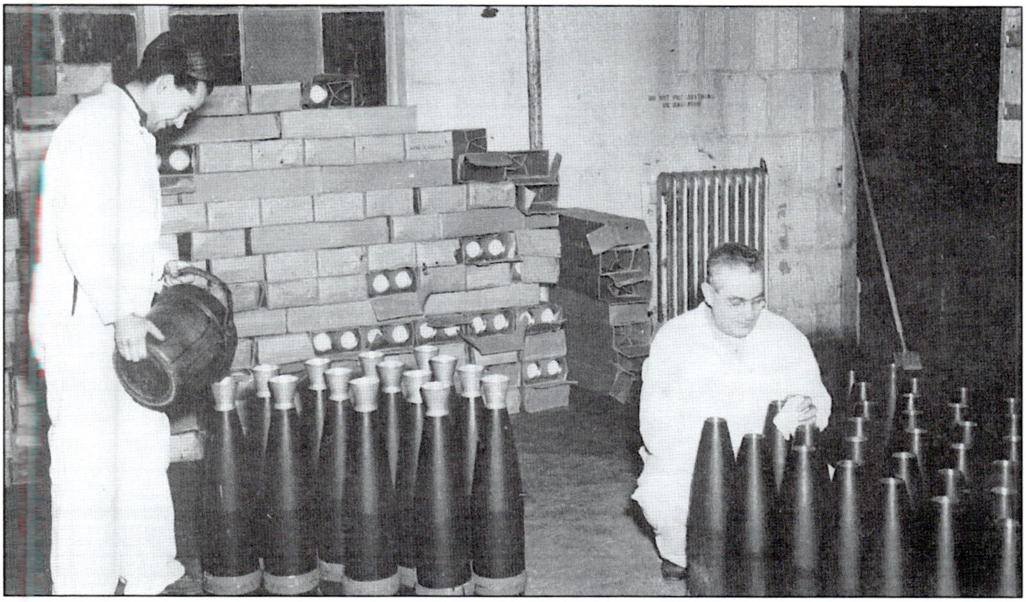

Pouring Explosives into Shell Casings. A worker pours liquid explosives ranging from Amatol to black powder, Trimonite, and TNT through funnels into shell casings while another cleans the shells. Loading operations started in warehouses in 1907. The new production lines, stressing safety and high efficiency, of rapid loading reached their peak during World War II. (U.S. Army Photograph-Picatinny Arsenal.)

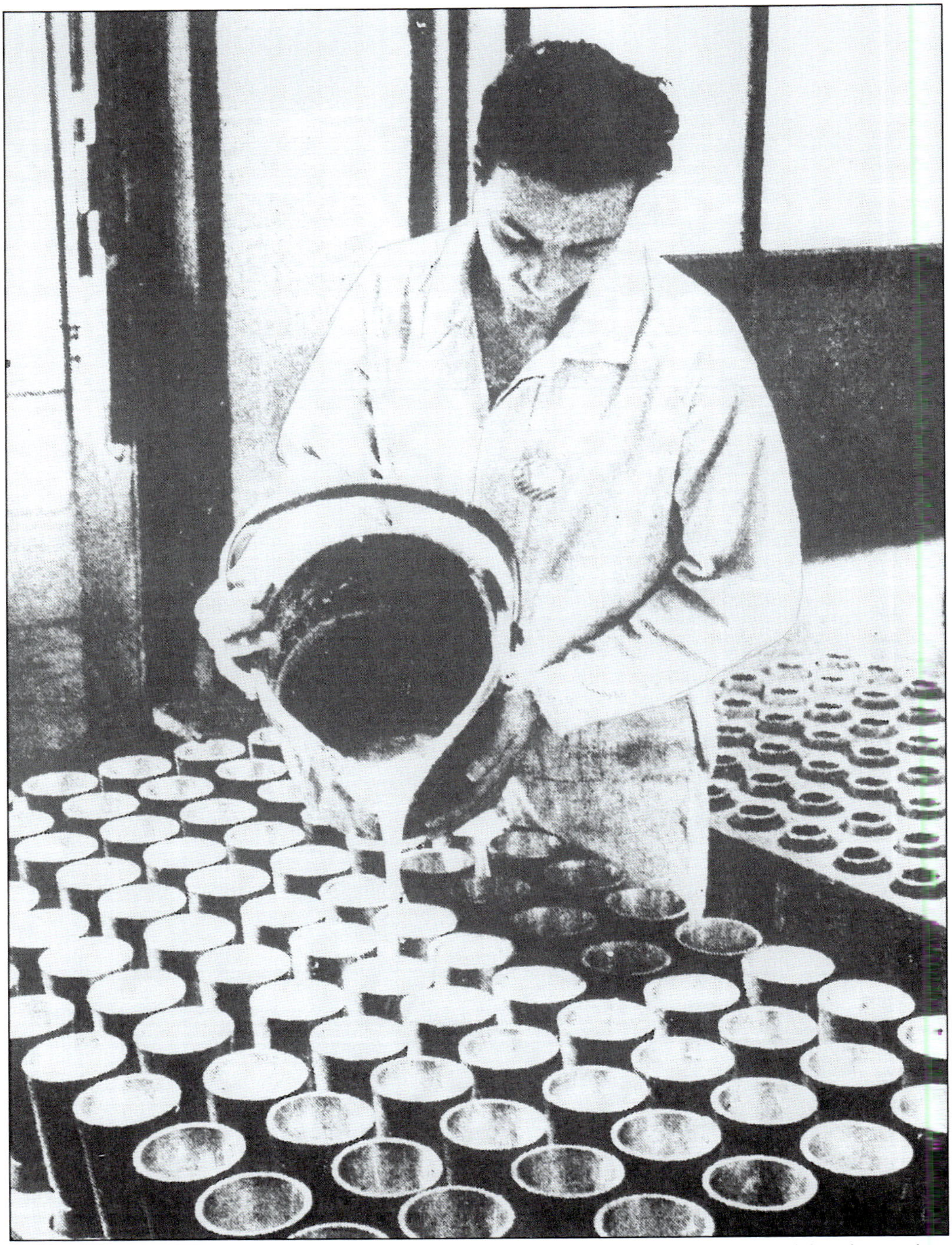

Pouring Liquefied TNT into Shell Casings. Melted in steam jacketed kettles with a capacity of 1,200 pounds each, heated TNT was drawn-off into galvanized iron cooling tubs. It was then transferred to the pouring room where it was manually stirred before being poured into shells from black iron or aluminum pails equipped with pouring spouts. (U.S. Army Photograph-Picatinny Arsenal.)

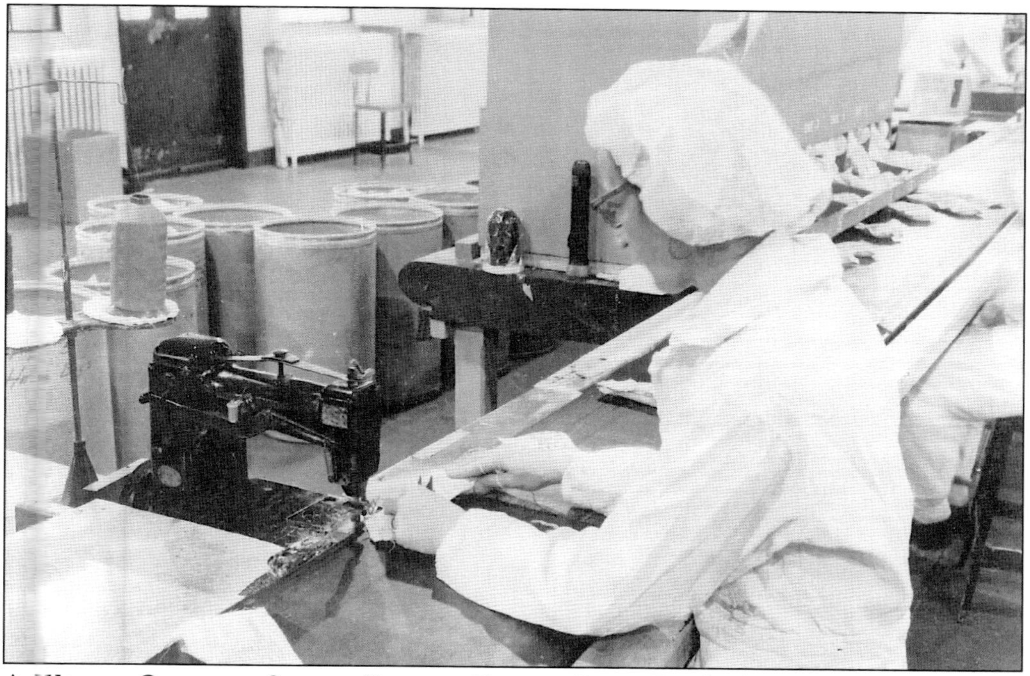

A WOMAN OPERATOR SEWING POWDER CHARGE BAGS. Bag loading of propellant powder began before the Spanish-American War and continued until the Vietnam conflict. Different types of guns required different amounts of powder, bagged in various sizes. The 16-inch guns of battleships and coast defense guns, for example, required 800 pounds of powder, the bags of which were 8 feet long. (U.S. Army Photograph-Picatinny Arsenal.)

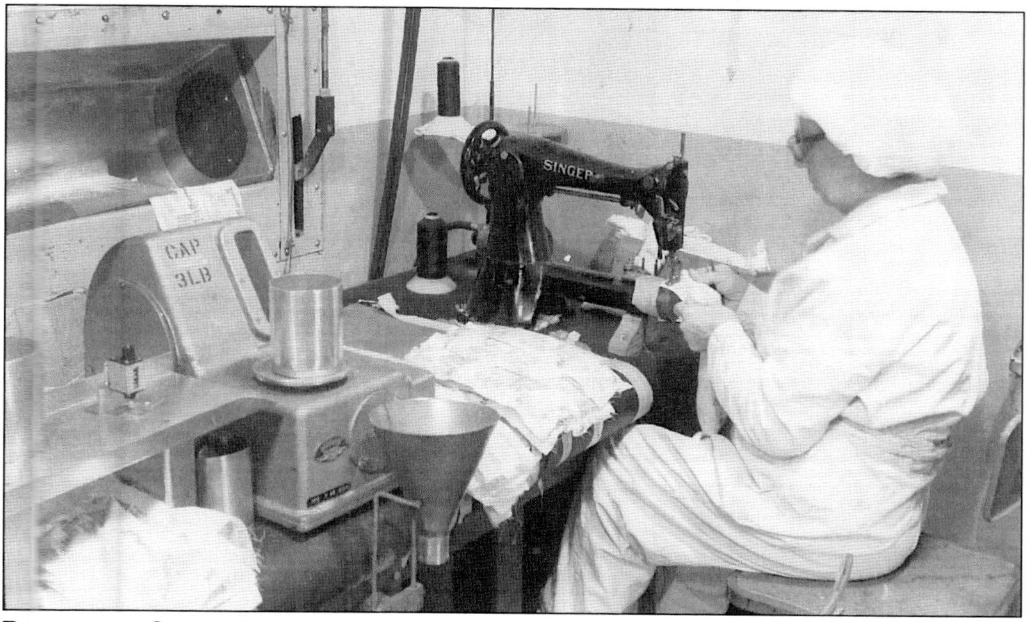

ROUND-THE-CLOCK OPERATIONS. Women gowned in special safety overalls and head gear sews a propellant powder bag at one of 500 sewing machines operated at arsenal production lines on a three-shift basis. In case of an emergency or rush order, local area women assisted by sewing bags at home. (U.S. Army Photograph-Picatinny Arsenal.)

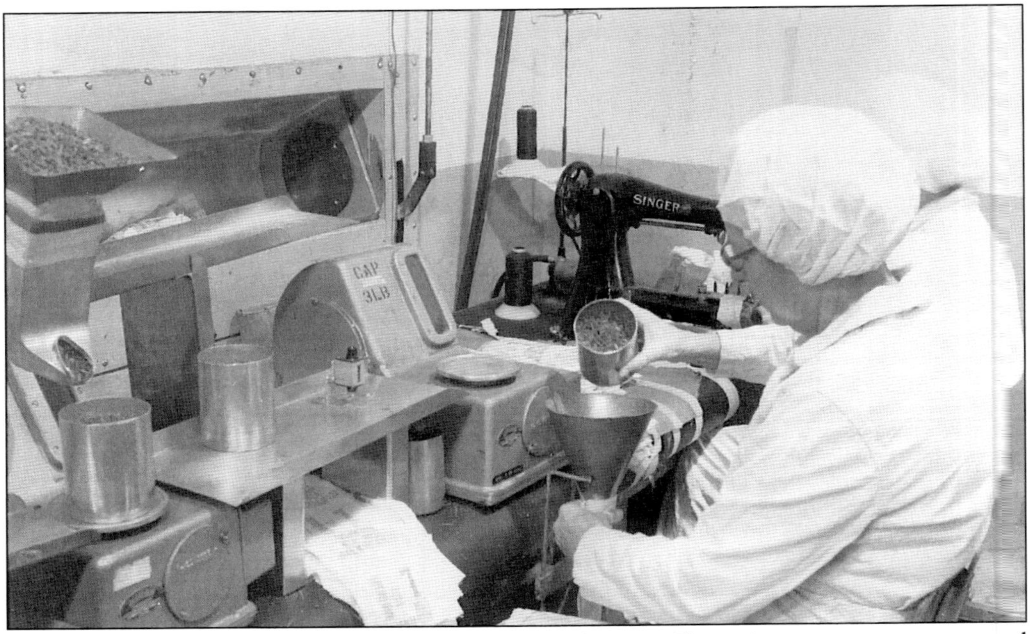

A WOMAN POURS PROPELLANT POWDER INTO A CHARGE BAG. A woman pours measured amounts of granulated powder into a small charge bag. In the background, another woman sews the bags. It was only after the arsenal was rebuilt after the 1926 explosion that safety features were considered, and bag-loading activities were placed in segregated and barricaded areas. (U.S. Army Photograph-Picatinny Arsenal.)

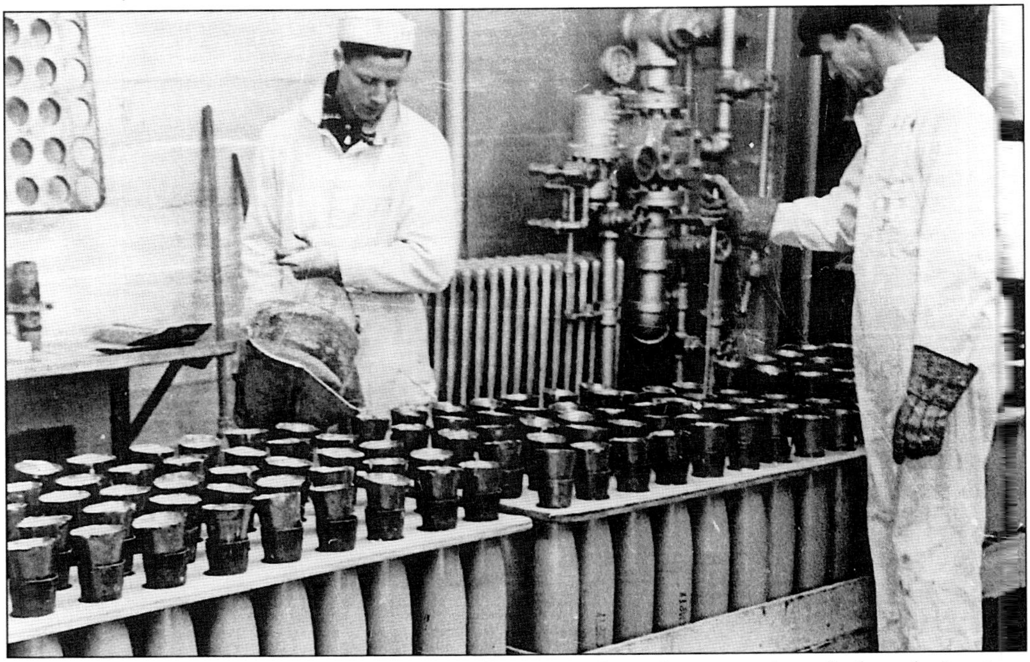

POURING LIQUID EXPLOSIVE INTO SHELL-CASINGS. A worker pours liquefied explosive into shell casings through a funnel supported by a wooden frame in the melt loading area. Another worker eases a rod into a filled casing to ensure even distribution of the explosive. The casings, all stenciled "empty," are on lift truck platforms. (U.S. Army Photograph-Picatinny Arsenal.)

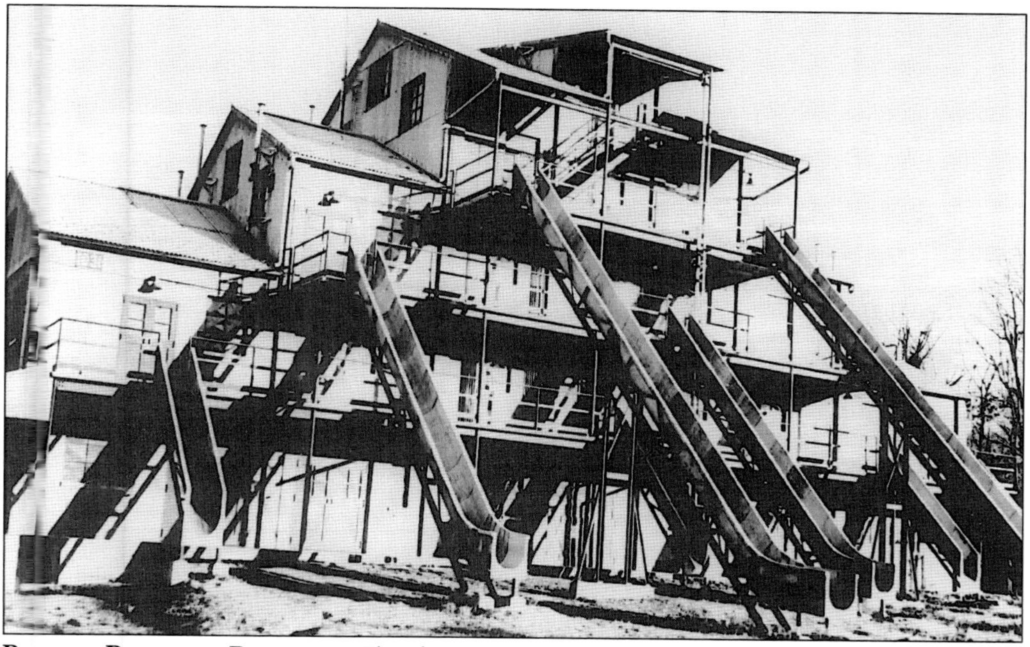

POWDER BLENDING BUILDING. This four-story powder blending building, with safety chutes, had sand at its base to soften landings. The powder, taken to the building's top in buggies, was dumped into a pit, where it fell through a hopper and funnel system for mixing, an operation that took 13 minutes. During the blending, no one was allowed in the building. (U.S. Army Photograph-Picatinny Arsenal.)

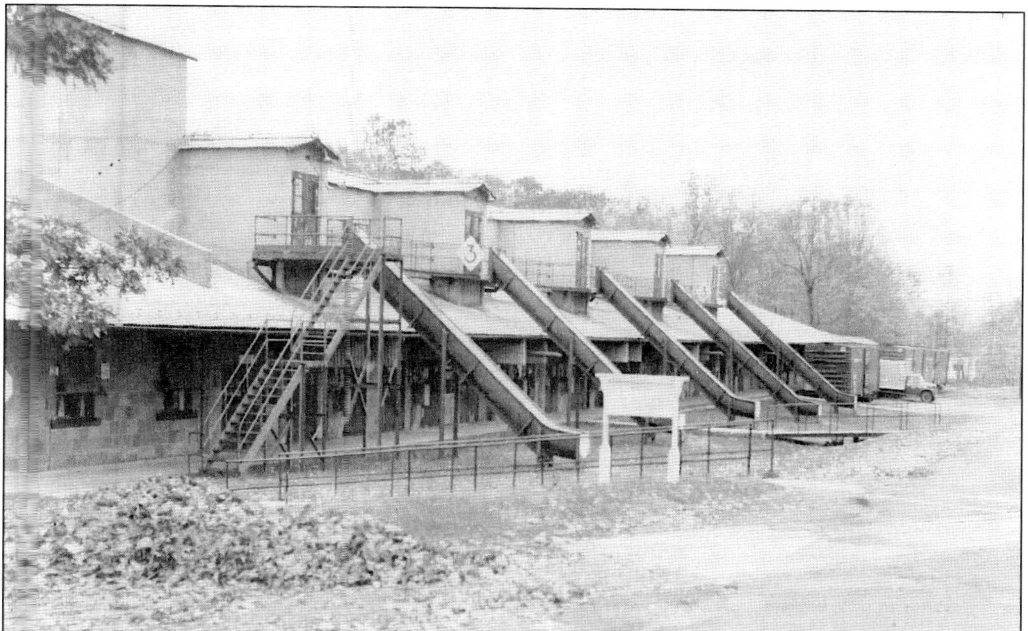

LOADING ASSEMBLY PRODUCTION. This and other assembly lines produced the bulk of ammunition from the bombing of Pearl Harbor until 1943, reaching a peak in the summer of 1942. In 24 hours, 40,000 37-mm, 30,000 60-mm, 7,500 8-mm, and 12,000 75-mm shells were assembled and loaded. (U.S. Army Photograph-Picatinny Arsenal.)

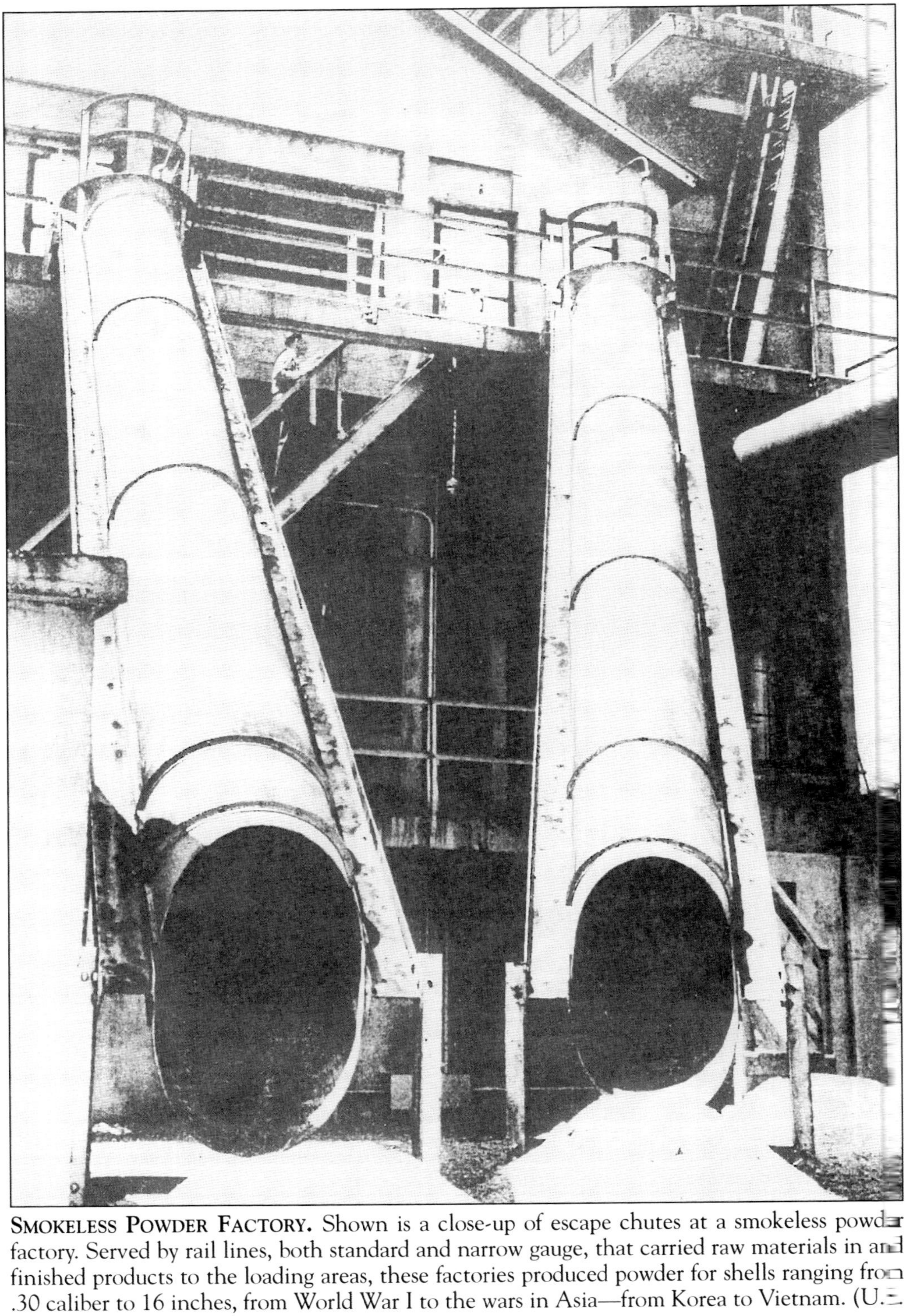

Smokeless Powder Factory. Shown is a close-up of escape chutes at a smokeless powder factory. Served by rail lines, both standard and narrow gauge, that carried raw materials in and finished products to the loading areas, these factories produced powder for shells ranging from .30 caliber to 16 inches, from World War I to the wars in Asia—from Korea to Vietnam. (U.S. Army Photograph-Picatinny Arsenal.)

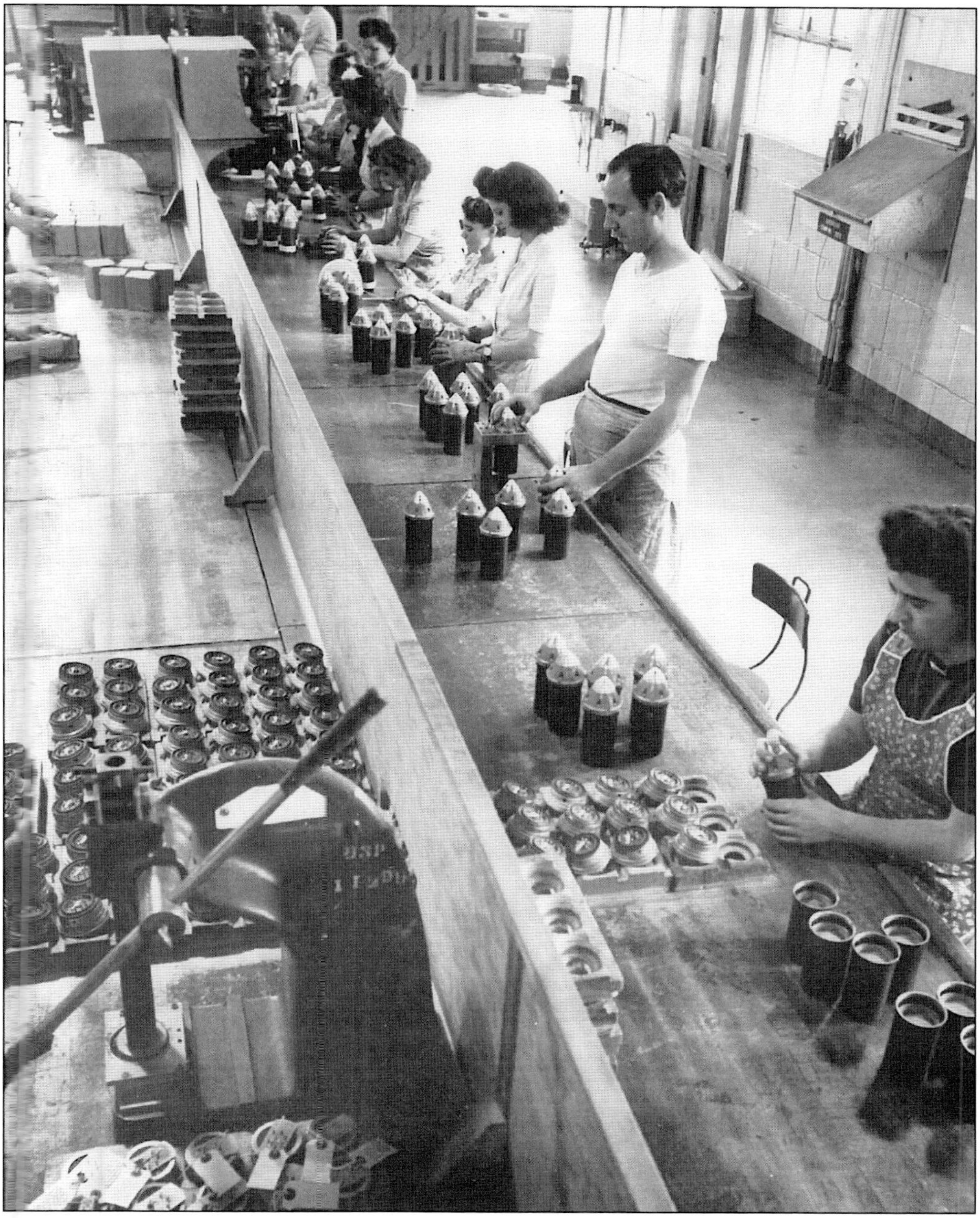

ROCKET FUSE ASSEMBLY LINE. Engineers working on fuses were forced to change the design and purpose as the character of World War II progressed. On-demand service became standard. For example, the navy's call for 200 special bomb fuses in 1944 resulted in a 48-hour period for loading, assembly, and delivery. (U.S. Army Photograph-Picatinny Arsenal.)

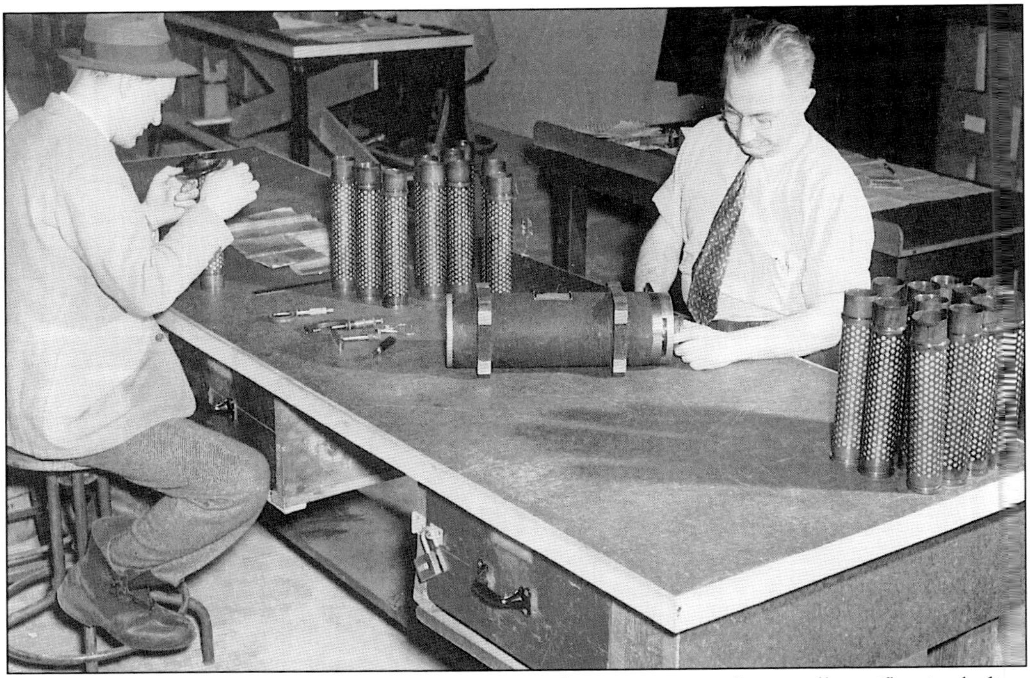

Recoilless Rifle Cartridge Case Inspection. The ammunition for recoilless rifles, including high explosives, high explosive antitank, smoke, and canister rounds plus new type fuses were developed and manufactured on arsenal production lines. A unique feature was the point-initiated shaped charge projectile for the 57-mm rifle. (U.S. Army Photograph-Picatinny Arsenal.)

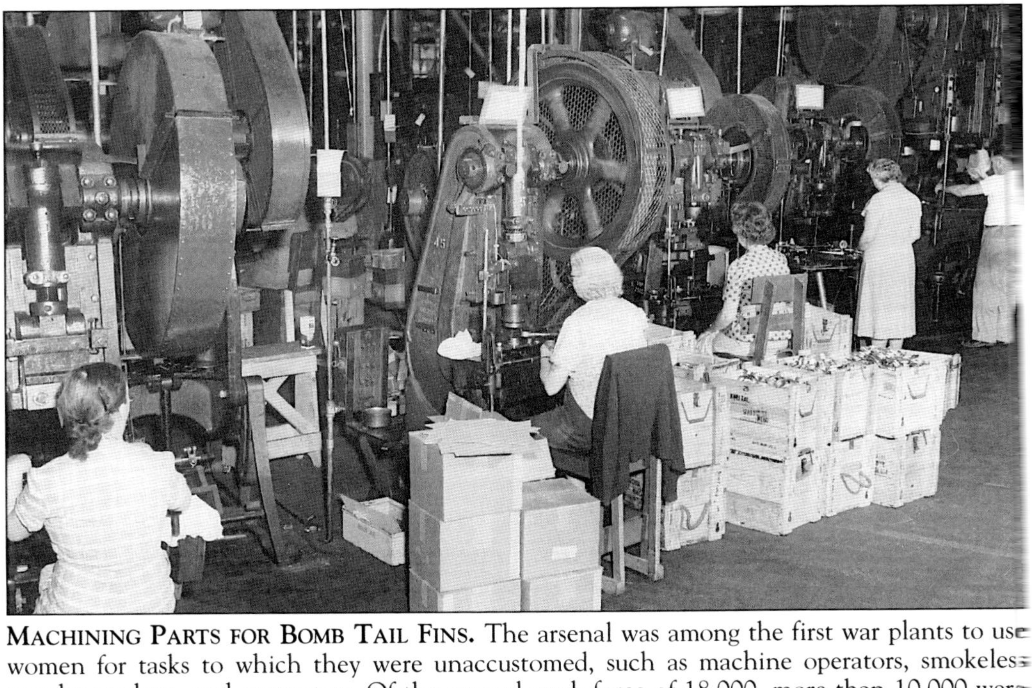

Machining Parts for Bomb Tail Fins. The arsenal was among the first war plants to use women for tasks to which they were unaccustomed, such as machine operators, smokeless powder workers, and carpenters. Of the arsenal work force of 18,000, more than 10,000 were women. (U.S. Army Photograph-Picatinny Arsenal.)

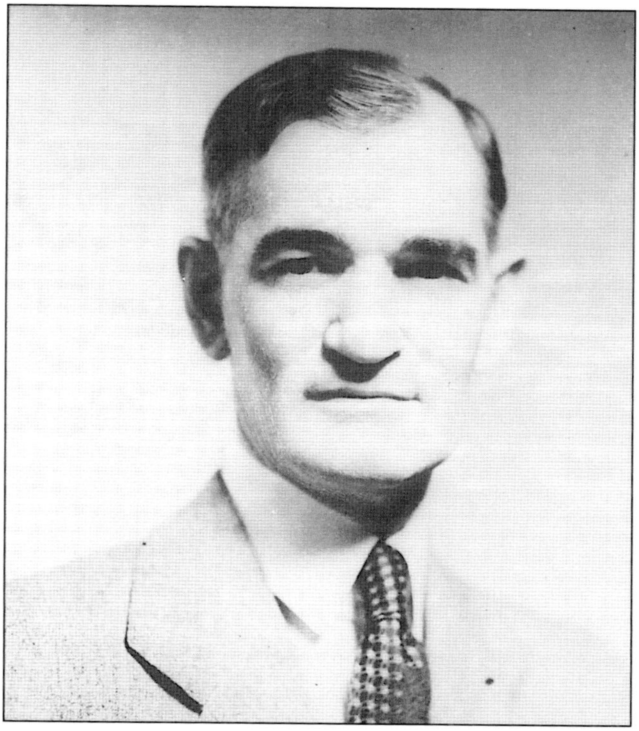

Dr. George C. Hale. He invented Haleite, a powerful explosive, patented in 1935. Earlier he developed a method to process cyclonite, which pierced armor plate TNT only dented. The army judged it too sensitive for military use. England, however, adopted it renaming it RXD. When the United States needed it for depth charges it had to get it from England by a form of reverse lend-lease. (U.S. Army Photograph–Picatinny Arsenal.)

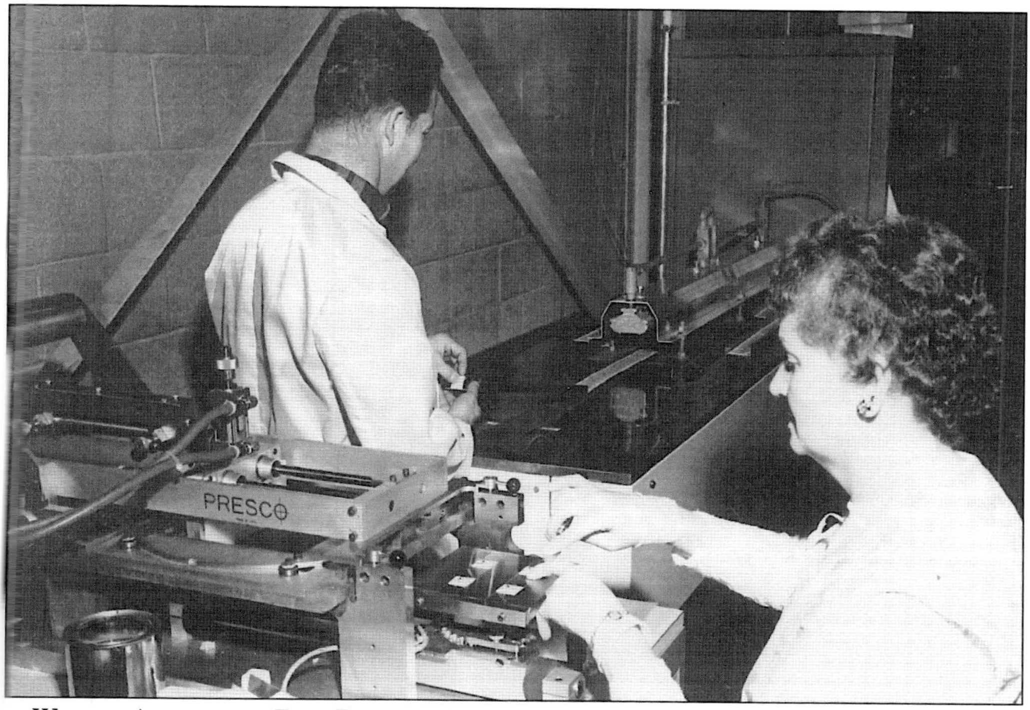

A Woman Assembling Fuse Parts by Machine. The arsenal's success in mechanization was highlighted by the tremendous jump in production of fuses, both special order and regular. In 1938, the total number produced by hand was 600. This figure rose to 173,000 in 1942. (U.S. Army Photograph–Picatinny Arsenal.)

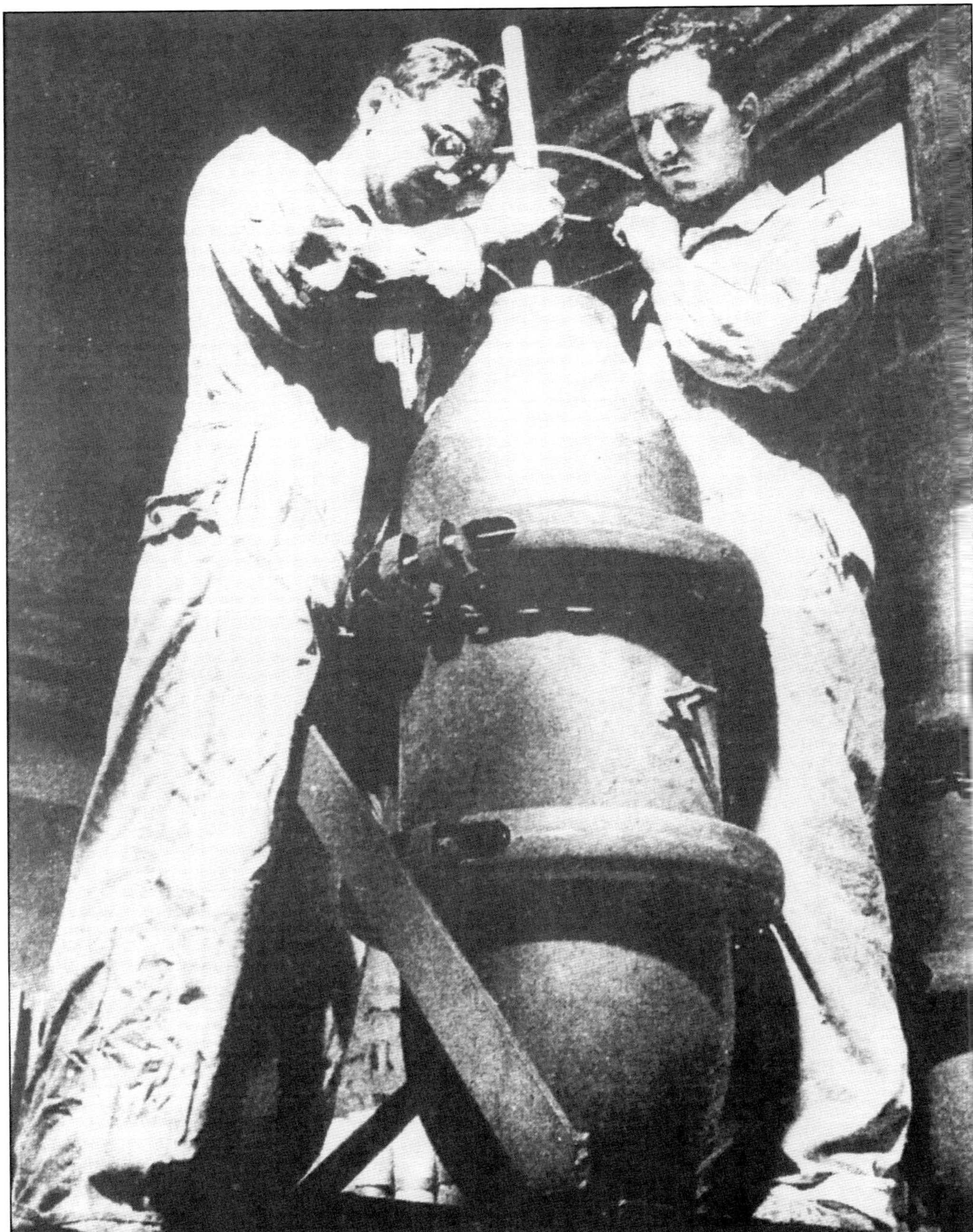

Tamping Down TNT in a 500-Pound Bomb. A combination of molten TNT and TNT pellets were used to fill 100-pound and larger bombs. Bombs and shells were not always filled in one pouring, depending upon the size and shape of the cavity for fear of shrinkage. In all cases it was necessary to break down the crust that formed, as pictured. (U.S. Army Photograph-Picatinny Arsenal.)

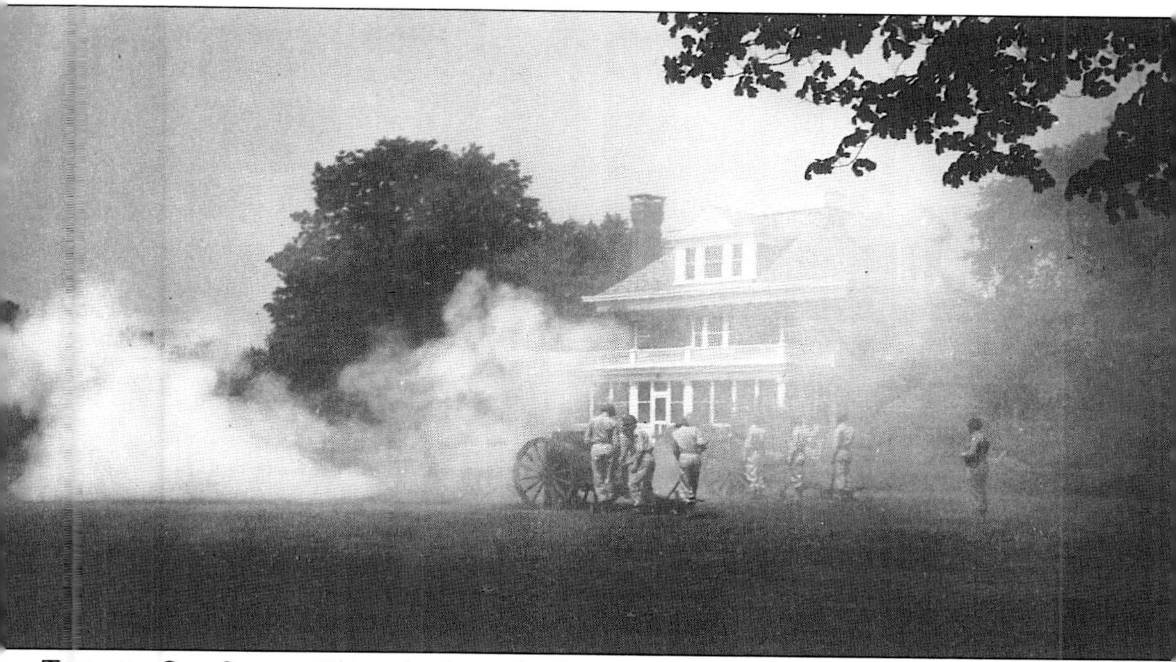

THIRTEEN GUN SALUTE. This is the first volly of a 13 gun salute, fired from a cannon on parade ground in front of the commanding officer's residence, as Gen. Everett S. Hughes, chief of the U.S. Army Ordnance Department, entered Picatinny Arsenal for "General Hughes Day," June 27, 1946. (U.S. Army Photograph-Picatinny Arsenal.)

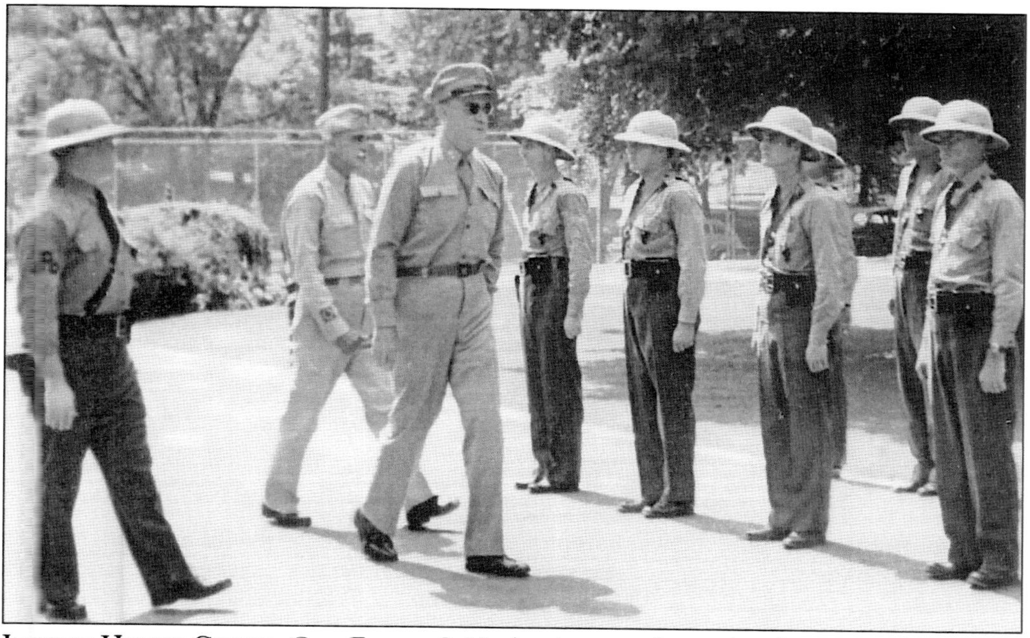

INSPECT HONOR GUARD. Gen. Everett S. Hughes inspects the civilian honor guard with Brig. Gen. William E. Larned (rear), the arsenal's war time commanding officer, as he returns to Picatinny for "General Hughes Day." He served as chief of the manufacturing group from 1935 to 1939. (U.S. Army Photograph-Picatinny Arsenal.)

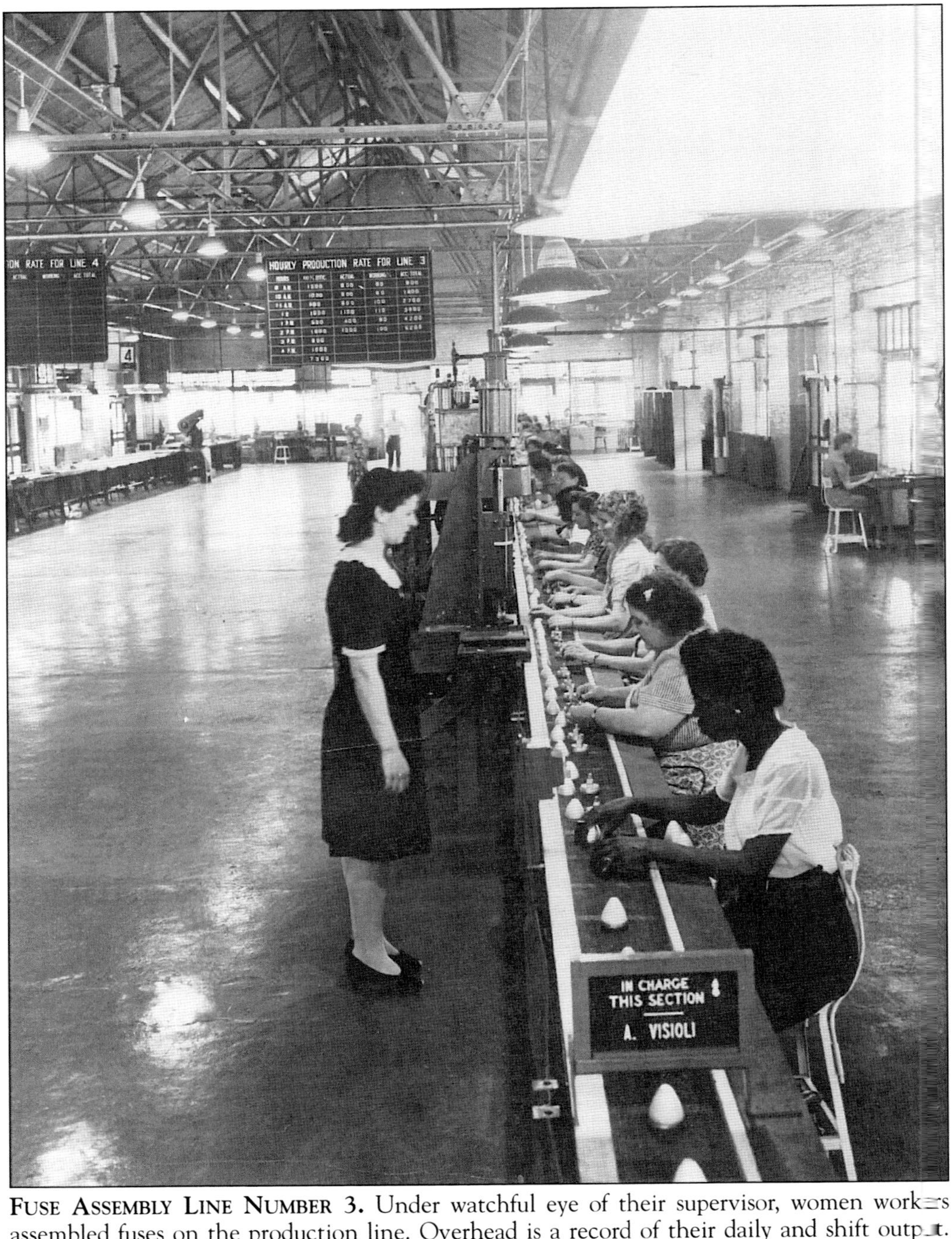

Fuse Assembly Line Number 3. Under watchful eye of their supervisor, women workers assembled fuses on the production line. Overhead is a record of their daily and shift output. Finished fuses were packed in metal lined boxes, soldered shut, checked for air leaks, painted, and marked. Finished fuses were fired statically to check for duds. (U.S. Army Photograph-Picatinny Arsenal.)

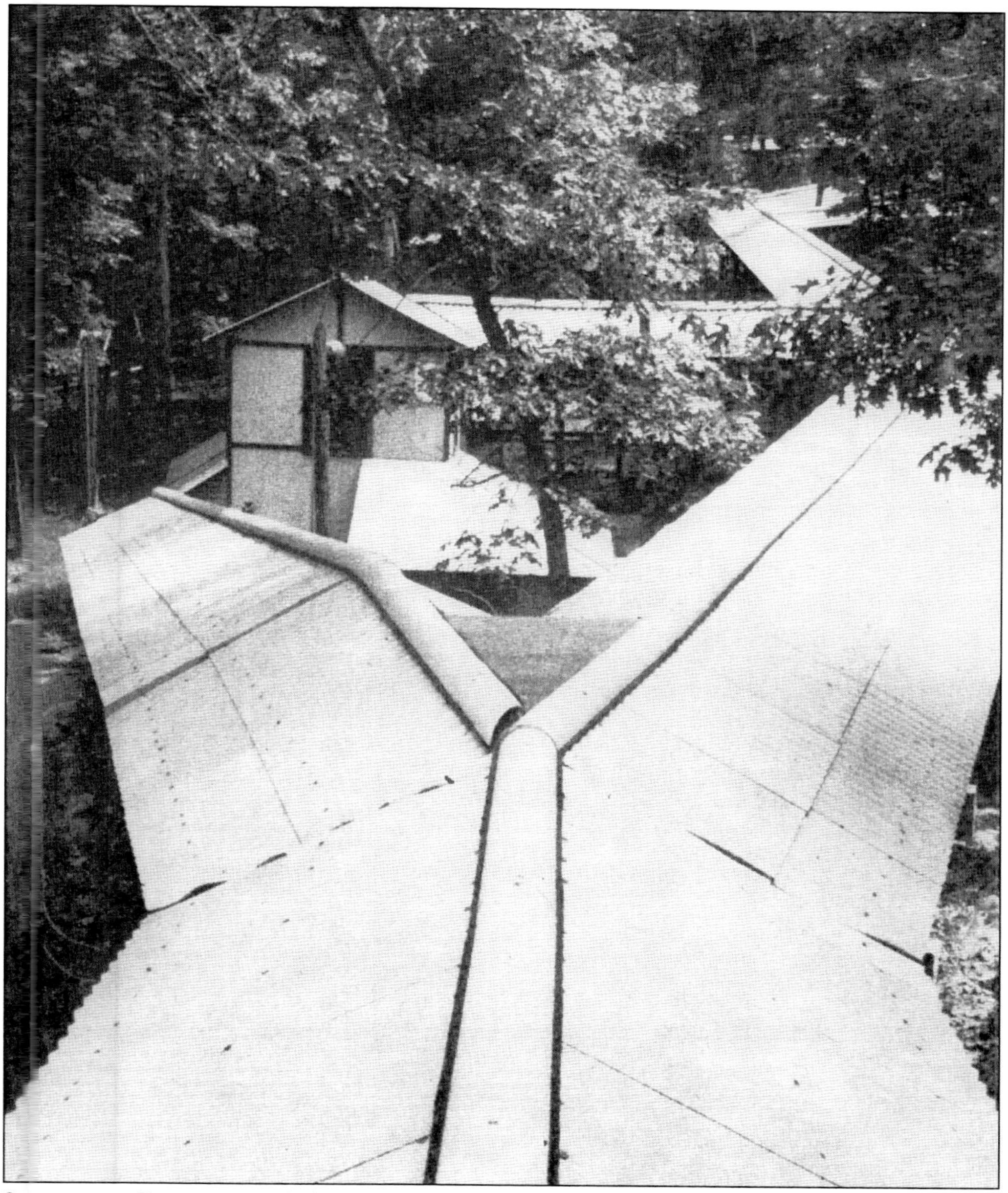

SHELTERED RUNWAYS. Sheltered runways connected small buildings, especially in long production lines in powder loading, pyrotechnic, and high explosives areas, to give workers sure footing from weather when in transit with dangerous materials. (U.S. Army Photograph-Picatinny Arsenal.)

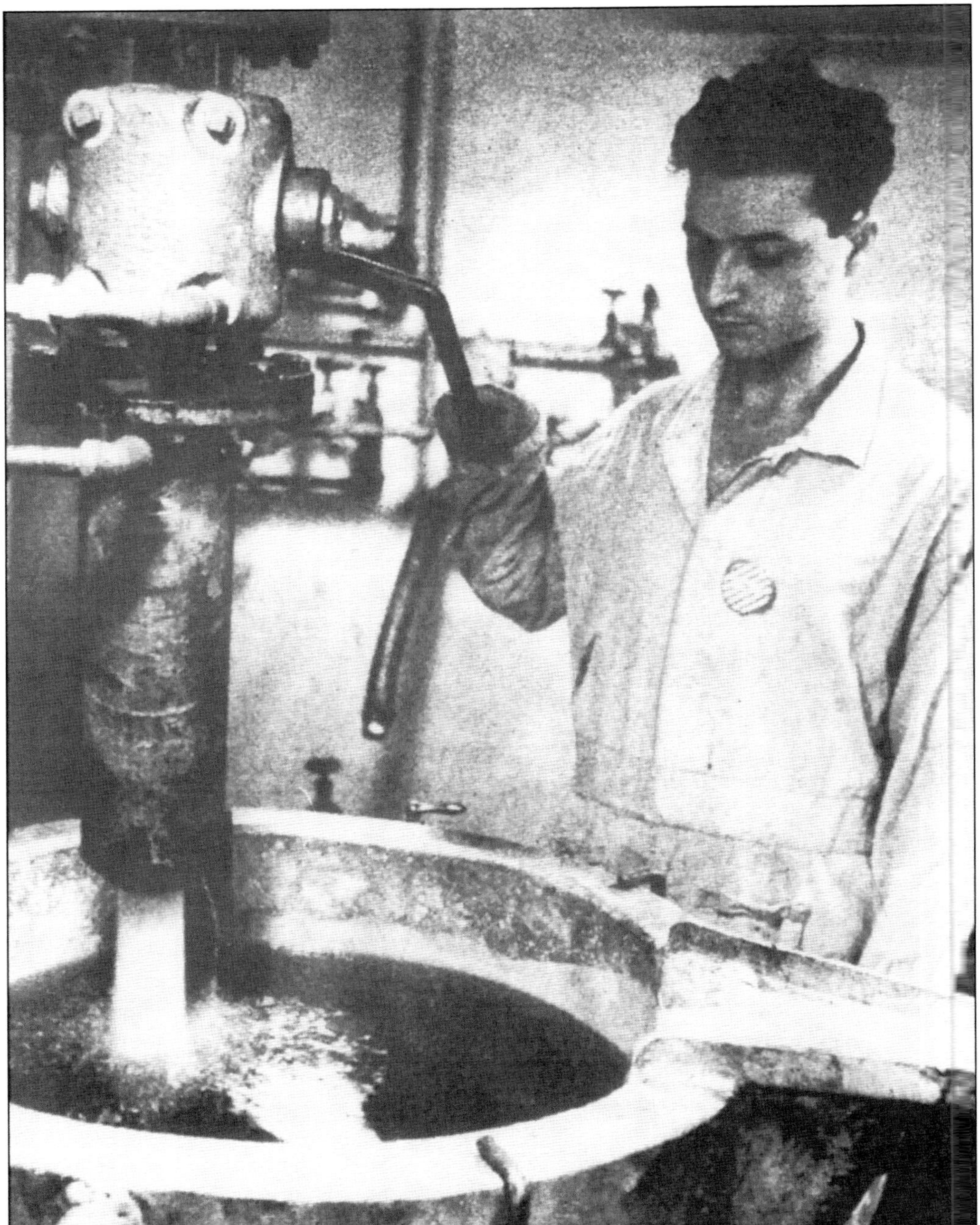

LOADING LARGE SHELLS WITH TNT. A workman wearing protective clothing, which had to be changed each shift, pours liquefied TNT into large shell casing by hand. TNT recovered from funnels was called clean scrap and re-melted. Floor sweepings were saved for re-crystallization. (U.S. Army Photograph-Picatinny Arsenal.)

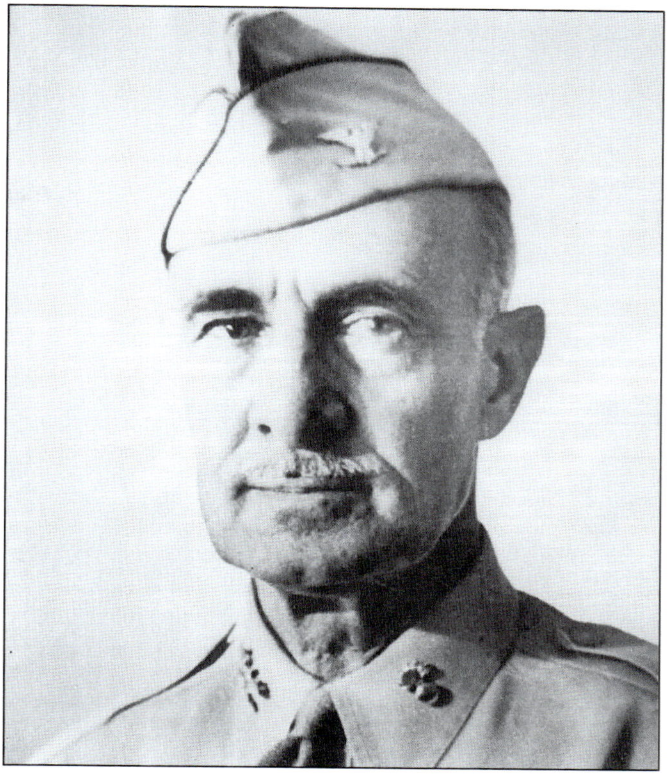

Brig. Gen. William Edmund Larned. The war time commander of Picatinny, Larned was promoted to brigadier general in 1942, two months after assuming command of the arsenal, where he had been chief of the Manufacturing Group since 1939. He was responsible for the overnight transformation of test pilot production lines into 24-hour-a-day production lines immediately after Pearl Harbor, and for the 6-fold increase in workers to 18,000. (U.S. Army Photograph-Picatinny Arsenal.)

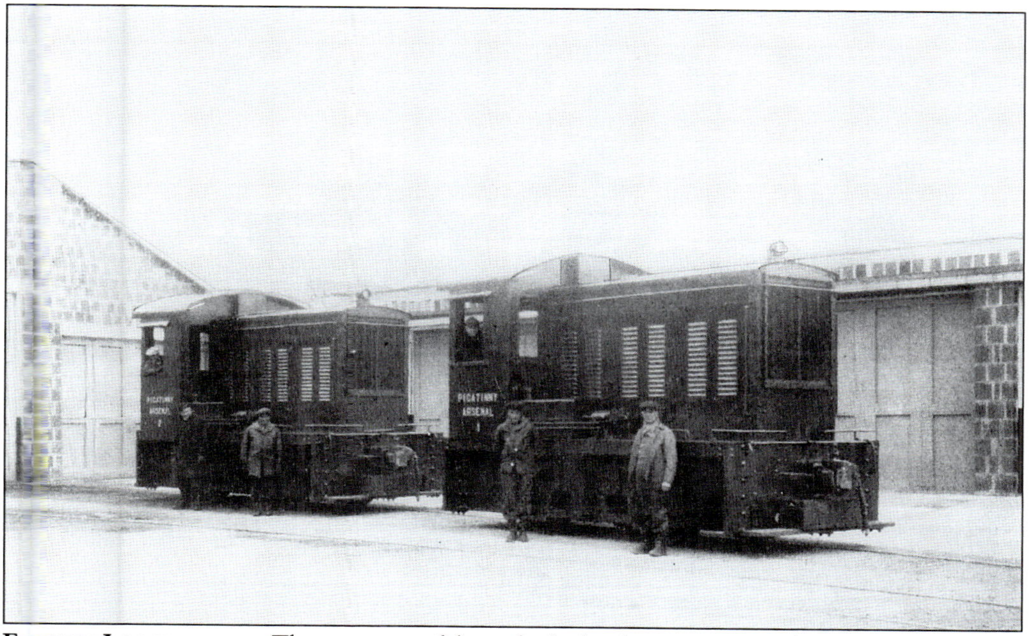

Fireless Locomotives. These are two of the eight fireless locomotives and crews that comprised the arsenal's rolling stock on its narrow gauge railroad, which connected many buildings, propellant loading, explosives, and testing areas. They hauled raw materials to a building then picked up the finished product for storage or shipment on the Wharton & Northern Railroad. (U.S. Army Photograph-Picatinny Arsenal.)

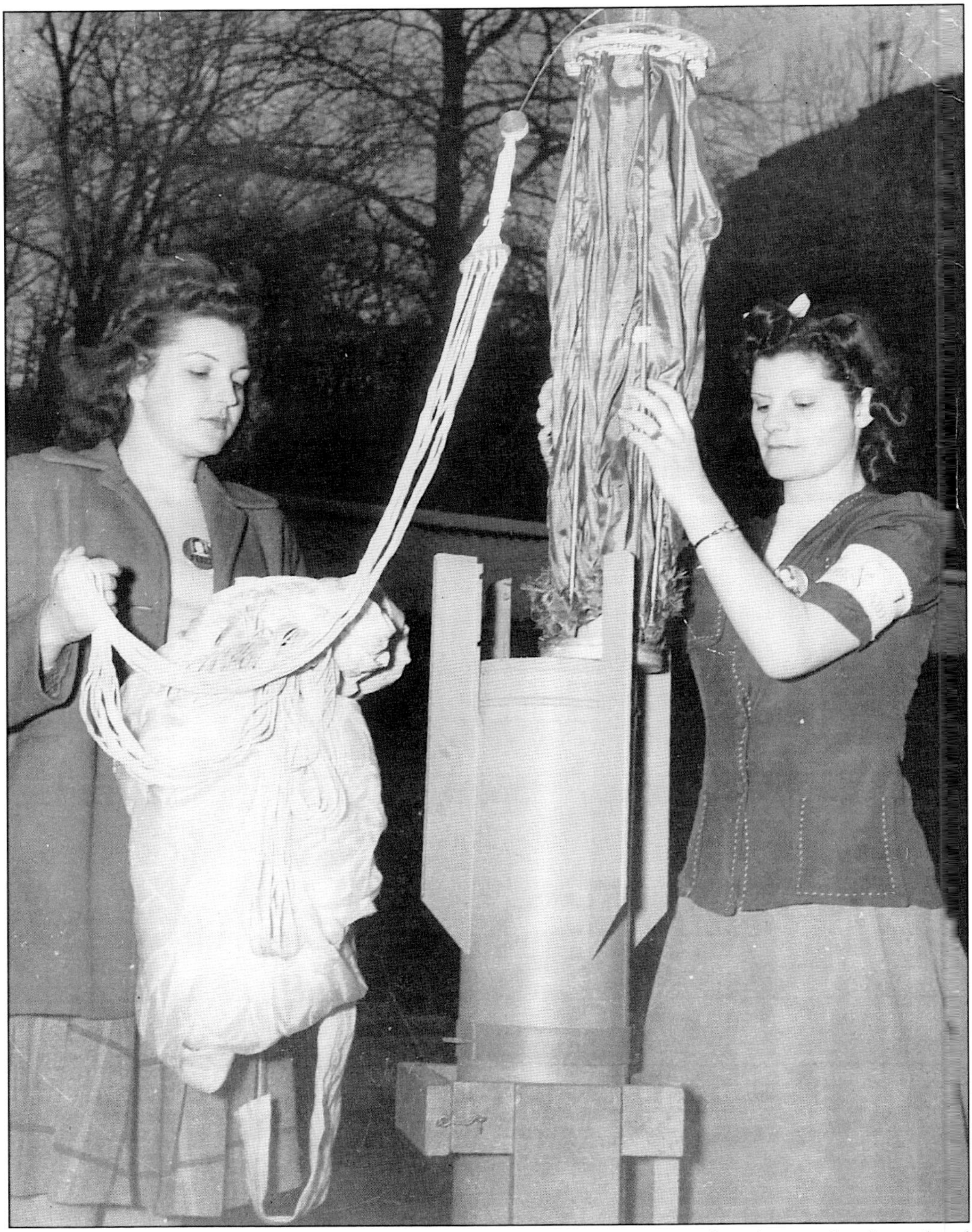

WOMEN LOAD PARACHUTE. A parachute is loaded into a canister bomb case for flares held upright in a wooden frame on June 5, 1943. At that time women numbered 48 percent of the work force totaling 18,000. In addition, the arsenal trained more than 15,000 key personnel from Picatinny, 12 Ordnance districts, 29 loading plants, 20 operating contractors, and Australia and Canada. (U.S. Army Photograph-Picatinny Arsenal.)

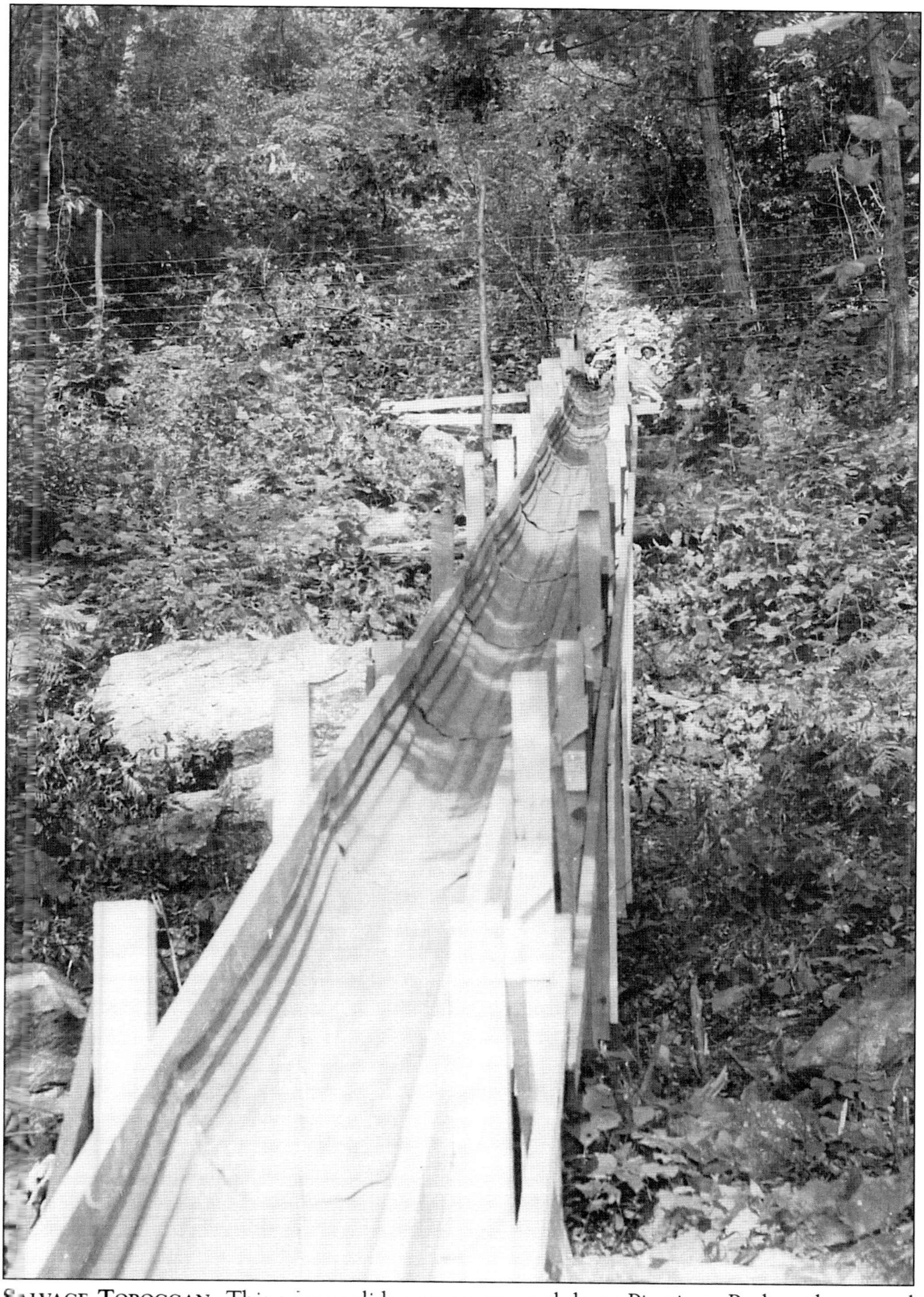

SALVAGE TOBOGGAN. This salvage slide was constructed down Picatinny Peak at the arsenal during World War II. Fragments of shell that were fired on the hill at the test range were collected, placed in a toboggan, and recovered at the base of the hill. They were then re-melted for re-use. (U.S. Army Photograph-Picatinny Arsenal.)

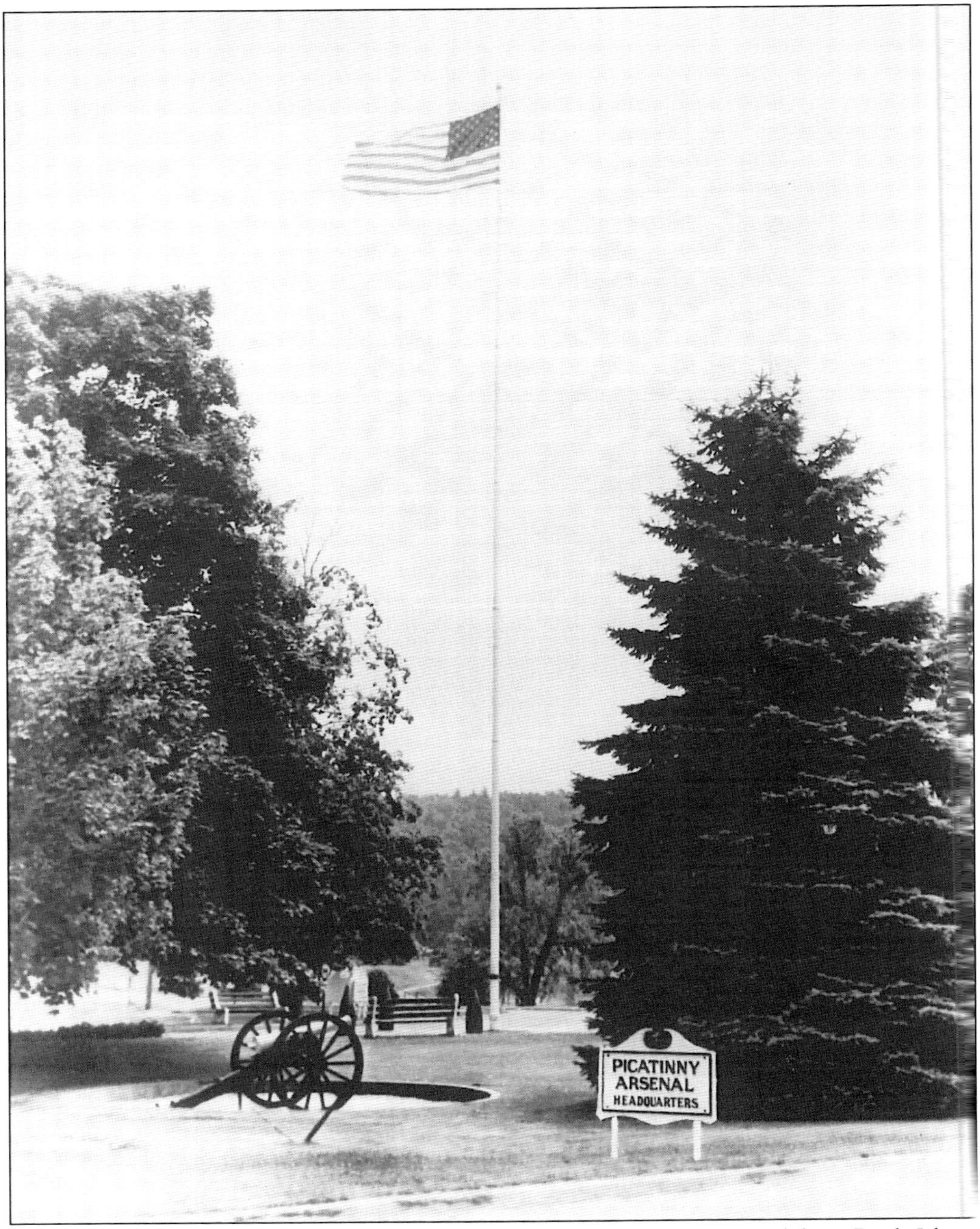

FLAG STAFF. This steel flag staff, flanked by two 6-pound cannon secured from Rock Islan[d] Arsenal, stood in the center of the junction of the roads in front of Building 151, the arsenal['s] main administration center from 1929 to 1981. When the new center was built, it was move[d] in front of it. (U.S. Army Photograph-Picatinny Arsenal.)

Five

THE POST WAR YEARS
EXPERIMENTATION AND DEVELOPMENT

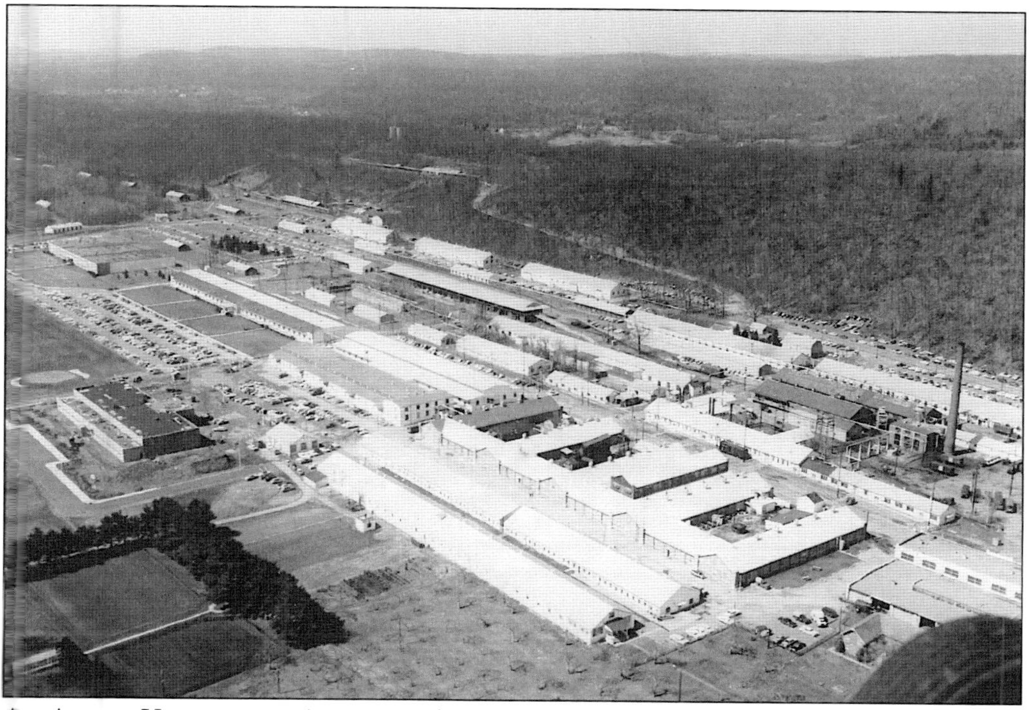

AN AERIAL VIEW OF THE ARSENAL. This view from northeast to southwest shows the lower area of the arsenal with workers' cars parked in lots. At the top is Navy Hill, site of the Naval Powder Depot that exploded in 1926. On the far left is a row of magazines. (U.S. Army Photograph-Picatinny Arsenal.)

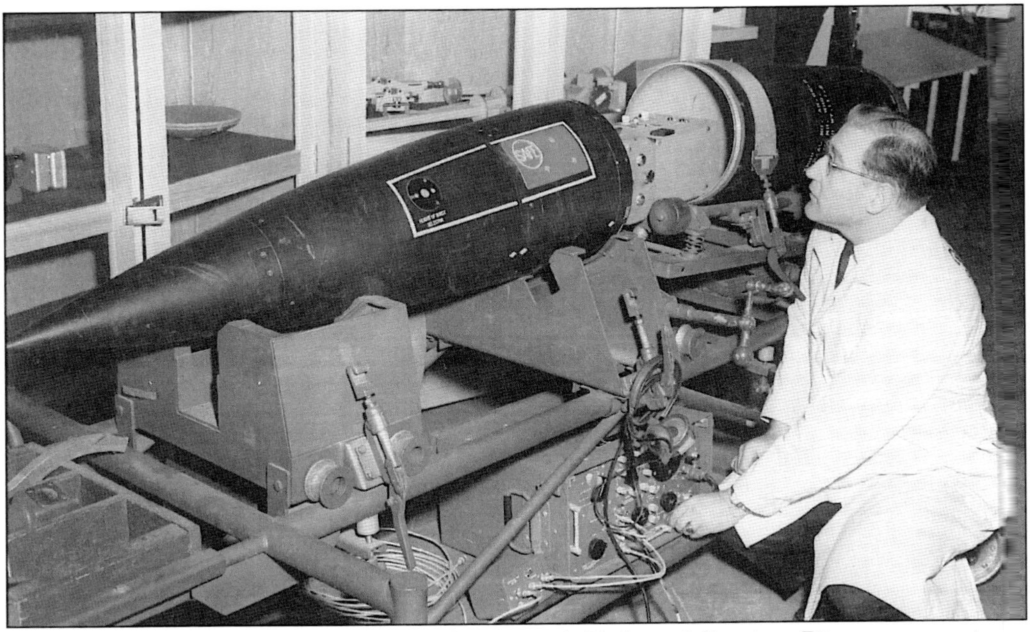

PREPARING NUCLEAR WARHEAD FOR TESTING. A Technical Services Directorate engineer prepares a nuclear missile warhead section for testing in the arsenal's radio frequency hazard simulation chamber. Other nuclear weapons perfected included a Universal Firing Device for a nuclear demolition munition, which provided as much blast effect as 500 boxcars of TNT. (U.S. Army Photograph-Picatinny Arsenal.)

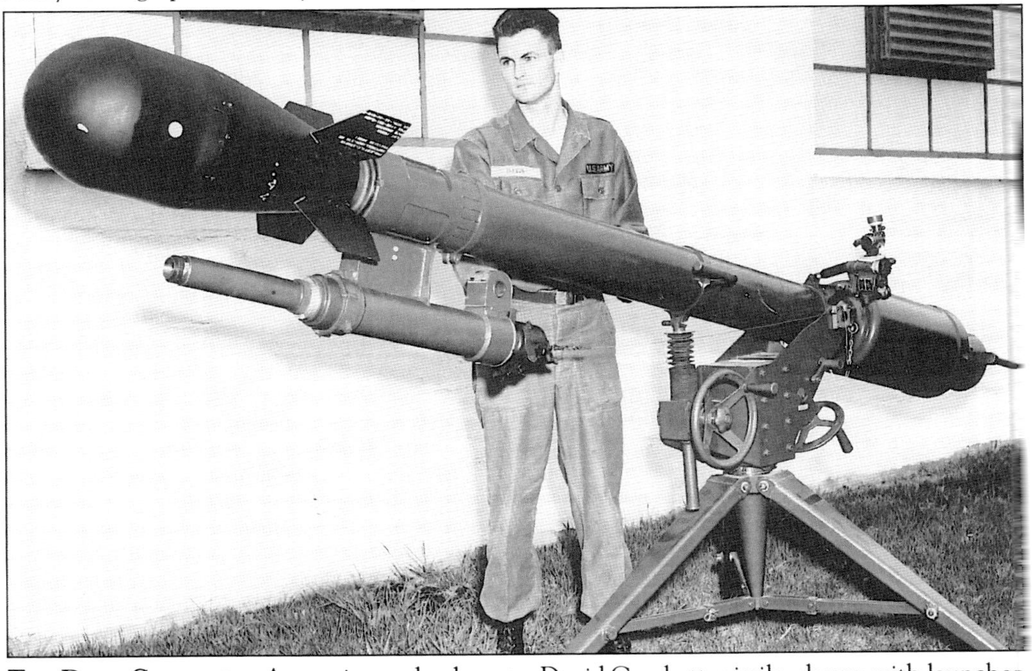

THE DAVY CROCKETT. An engineer checks out a David Crockett missile, shown with launcher, projectile, and spotting device. Developed for the infantrymen, it is carried and fired from a tripod by two or three men. A heavier version with longer range was mounted on a jeep or armored personnel carrier. (U.S. Army Photograph-Picatinny Arsenal.)

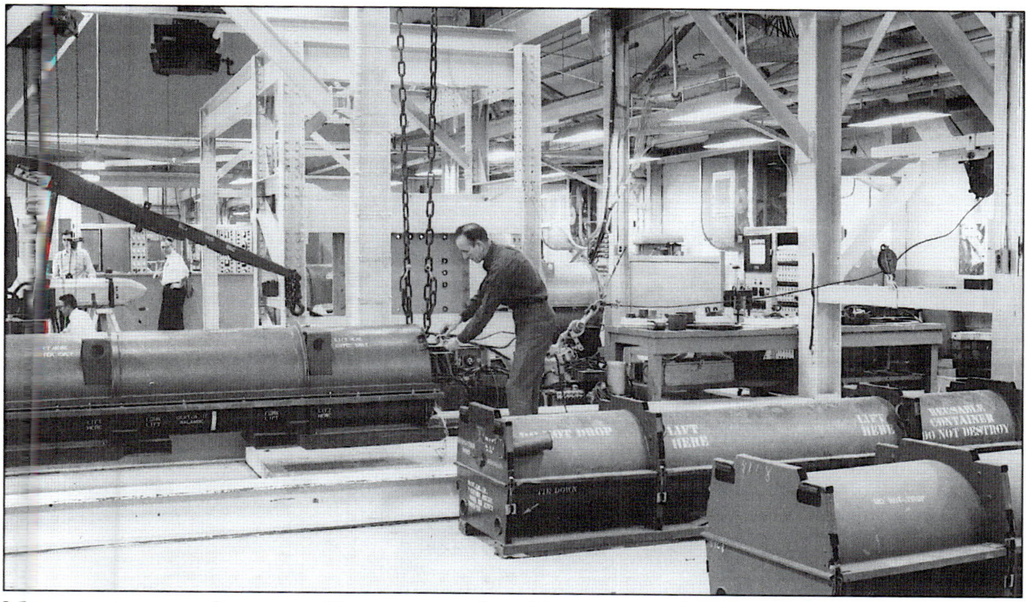

MANUFACTURING SHIPPING CONTAINERS. In addition to producing ammunition, the arsenal manufactured fiber containers, boxes, and bundle packs to insure safe delivery. Packaging engineers supplemented the arsenal's staff to develop better methods of packing bombs, explosives, and ammunition while engineers from private industry came to Picatinny to learn the science of packing. (U.S. Army Photograph-Picatinny Arsenal.)

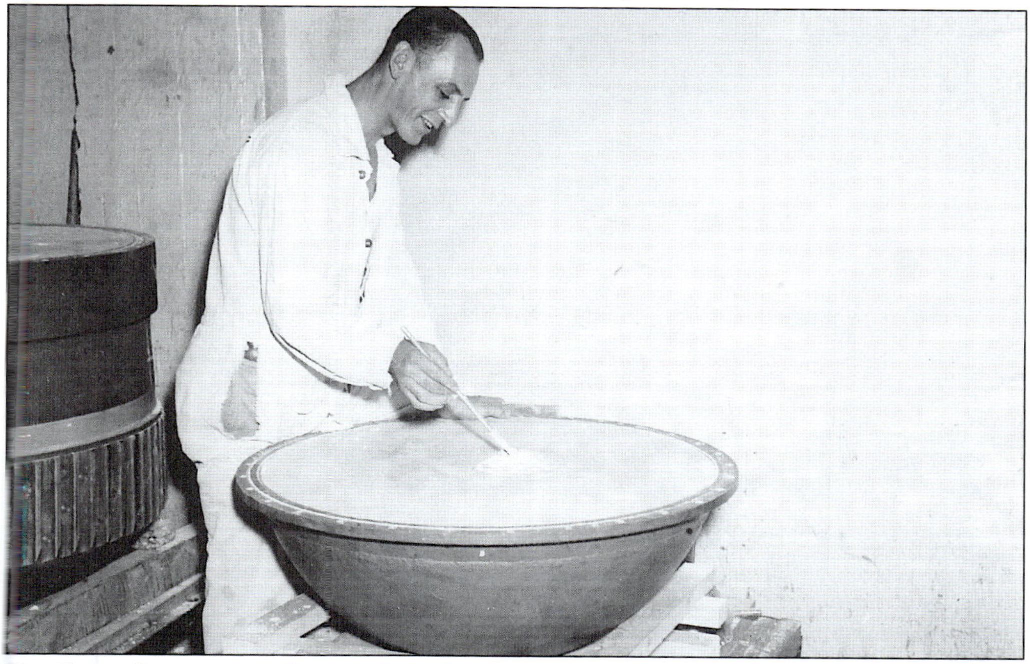

THE INERT LOADING OF BOMBS. A worker wearing protective clothing stirs solidifying liquid explosive in a T2 bomb head prior to final assembly of the bomb. Explosives ranged from 50 percent each of Amatol and TNT, to Trimonite. All bombs were painted with two coats of paint or shell lacquer and sealed with colored adhesive tape, which indicated the bomb type. (U.S. Army Photograph-Picatinny Arsenal.)

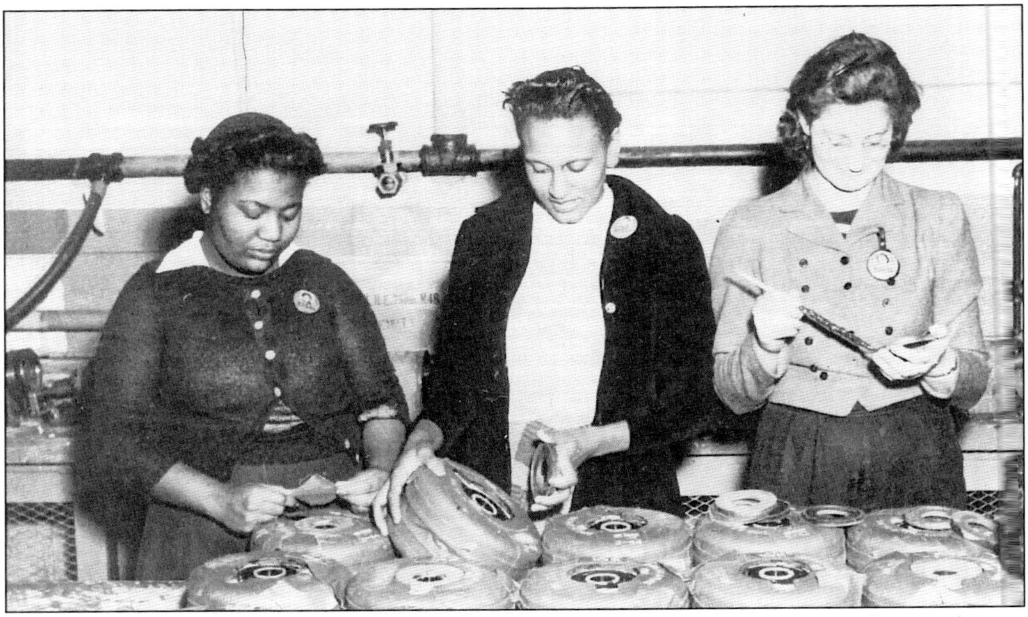

Assembling Mines. Women employees assemble M-5E3 land mines on this production line in 1944. This mine, thousands of which were produced at the arsenal for shipment to Europe and the Pacific theaters of war, was one of many modifications to the M-1 anti-tank mine. (U.S. Army Photograph-Picatinny Arsenal.)

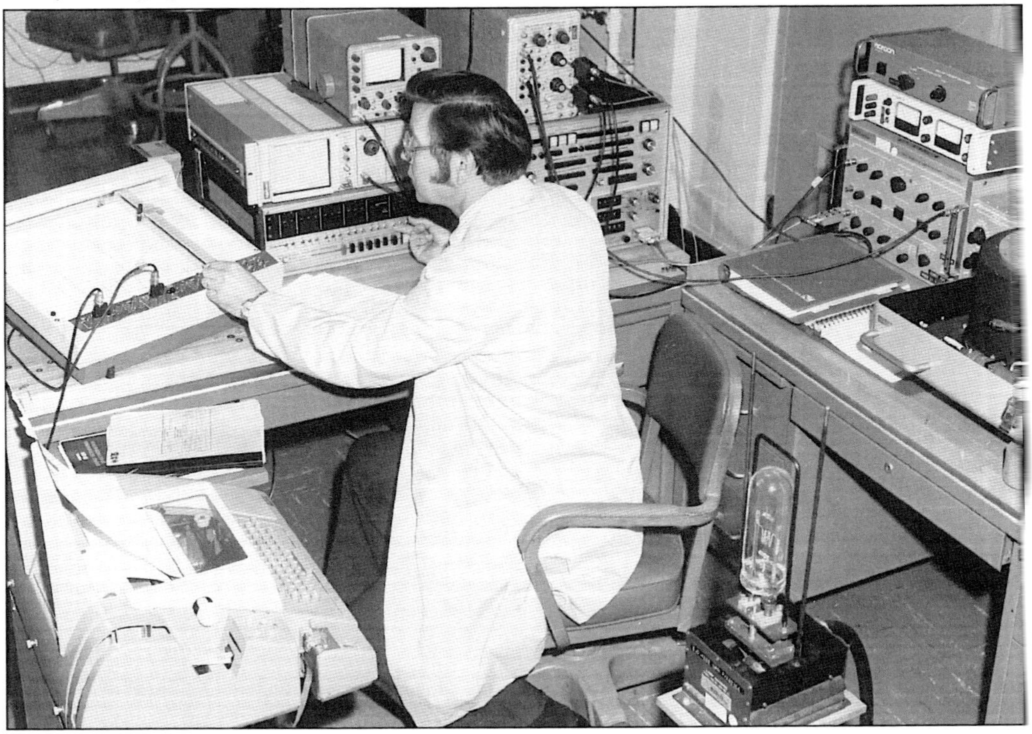

Studying Burning Furnaces. A technician operates a fast scan grating spectrometer with millisecond time resolution used to study burning surfaces of flares, propellants, and solid fuel used in rockets. (U.S. Army Photograph-Picatinny Arsenal.)

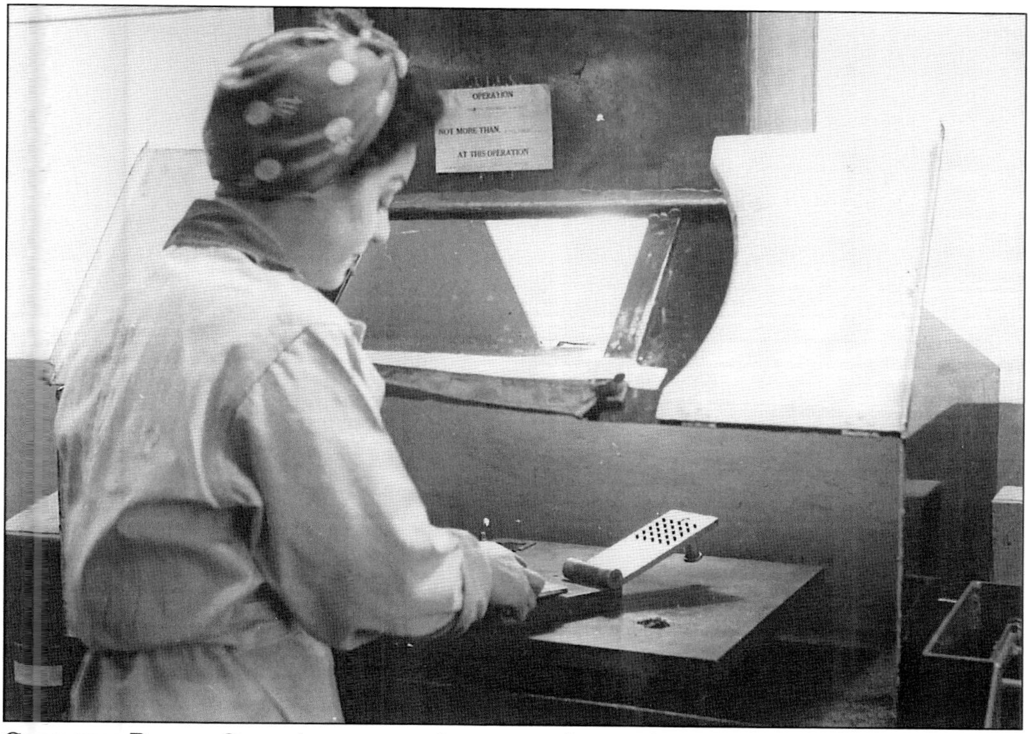

CHARGING PRIMER CAPS. A woman worker is seen charging M60 primer caps. The live primers, assembled in cartridge cases, then went to the powder weighing room to receive a charge, then to the assembling and crimping machine where the shell and cartridge were assembled together. Due to mechanization, the output of primers increased from 10,000 in 1938 to 90,000 in 1942. (U.S. Army Photograph-Picatinny Arsenal.)

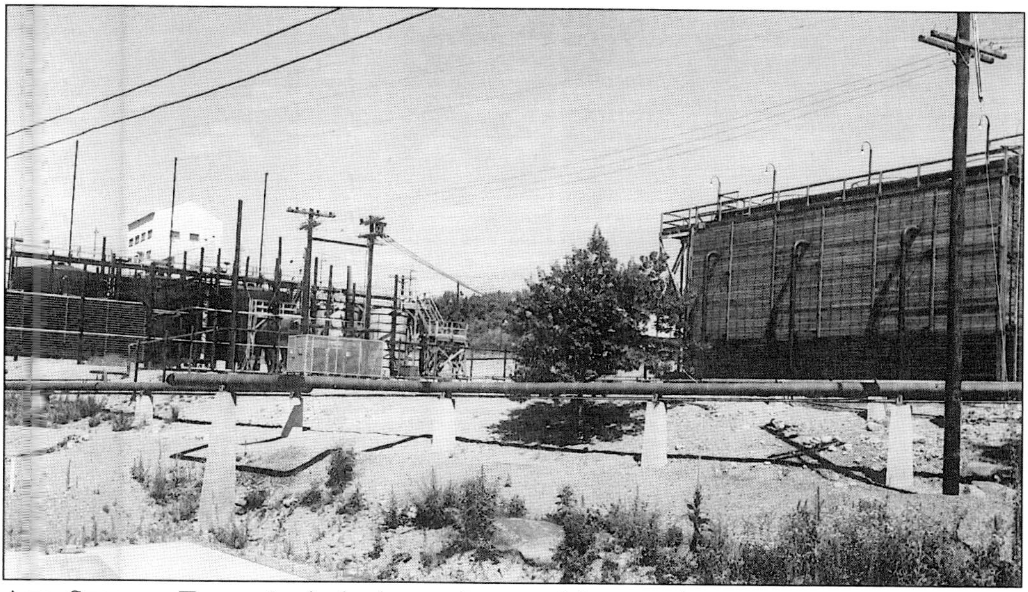

ACID STORAGE TANKS. In the background is one of the arsenal's powder factories. Pipes in the foreground carried the acid to the different buildings in which acid was used in development and manufacture of pilot lots of munitions. (U.S. Army Photograph-Picatinny Arsenal.)

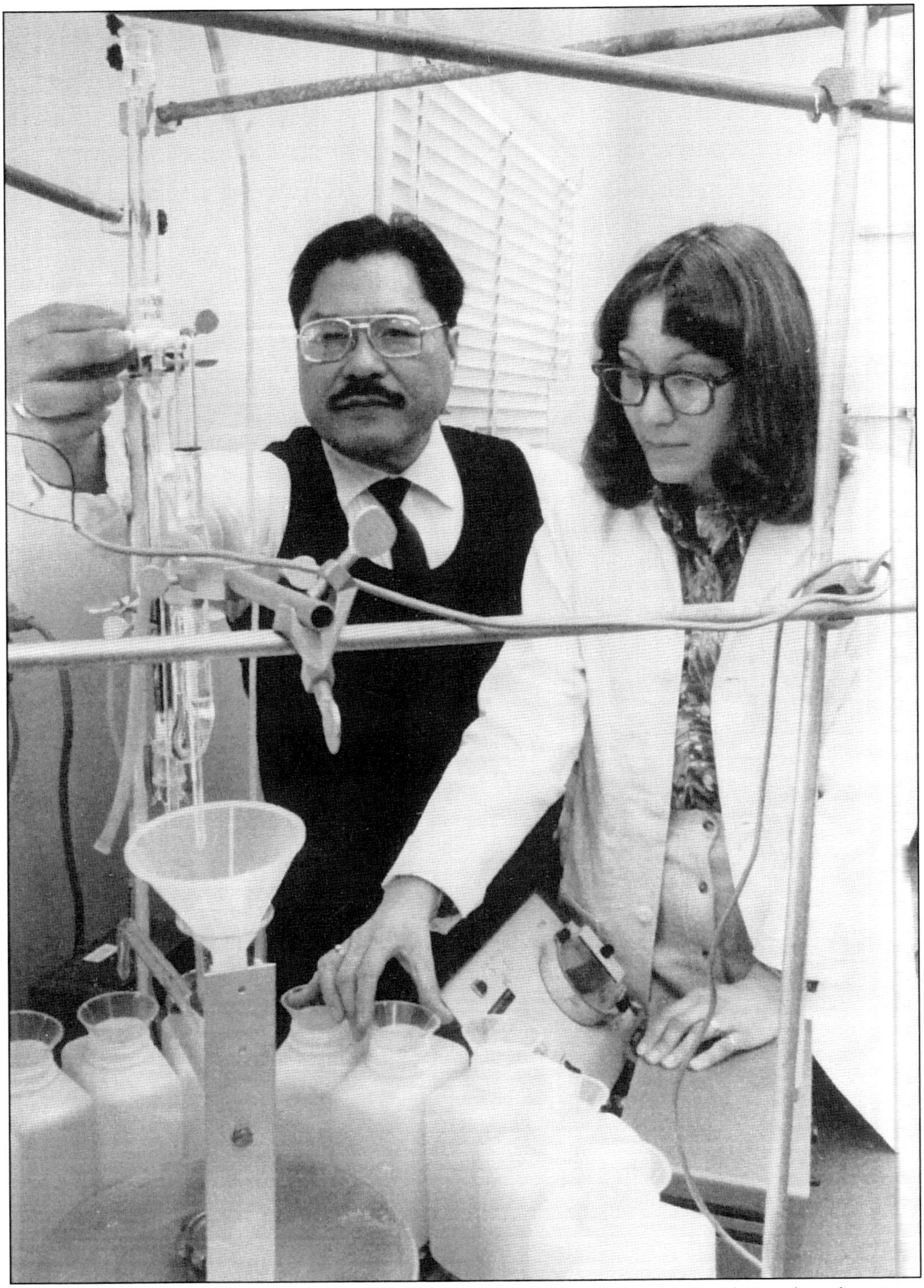

CHEMISTRY LABORATORY. Dr. Harold Matsuguma and an assistant work in the chemistry laboratory, checking the flow of chemicals placed in different containers for further examination. Built in 1941 and rehabilitated in 1980, the laboratory was the site of many of Picatinny's World War II research efforts. (U.S. Army Photograph-Picatinny Arsenal.)

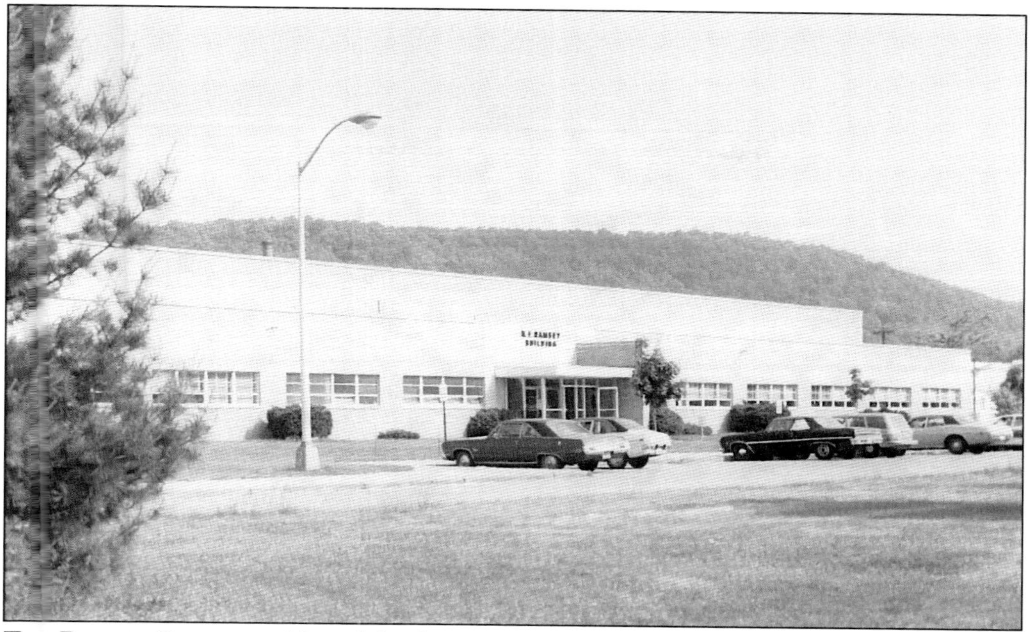

THE RAMSEY BUILDING. Named for Gen. Norman F. Ramsey, arsenal commander (1922-26) and winner of the 1989 Nobel Physics Prize, the Ramsey Building is now the Quality Engineering Directorate headquarters. He was cited for bravery in 1926 when, "with utter disregard for his personal safety he entered a burning building containing loaded bombs to aid in rescuing an injured worker." (U.S. Army Photograph-Picatinny Arsenal.)

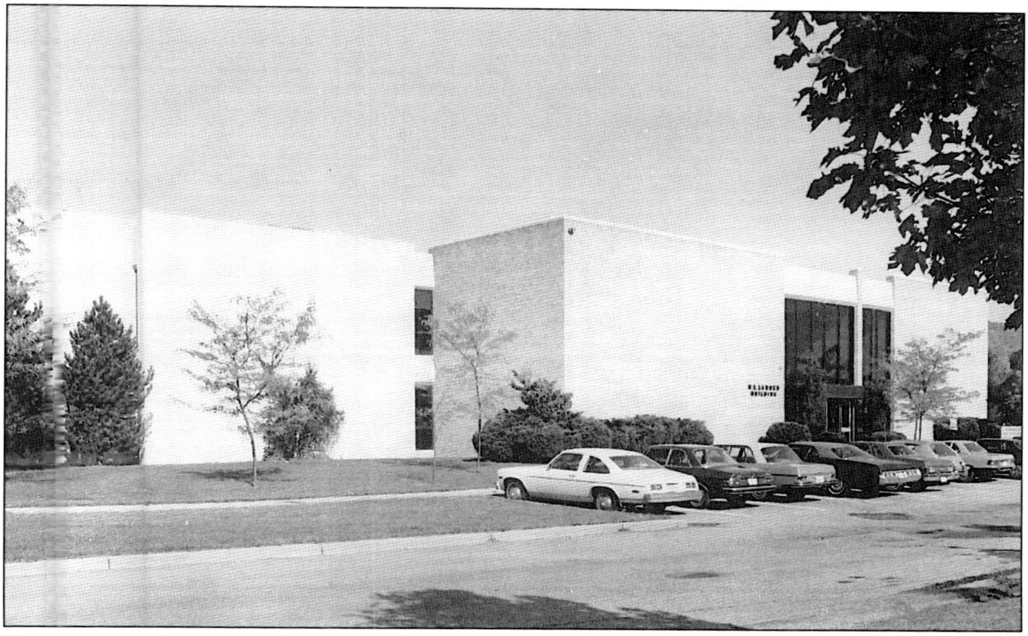

THE LARNED BUILDING. Named for Brigadier General William E. Larned, arsenal commander (1941–1948), the Larned Building is headquarters for the Fire Support Armament Systems Integration Division, Mines and Demolition Division, and Program Office. His contributions to mechanization, research, and development earned him the Legion of Merit and Army Commendation Medal. (U.S. Army Photograph-Picatinny Arsenal.)

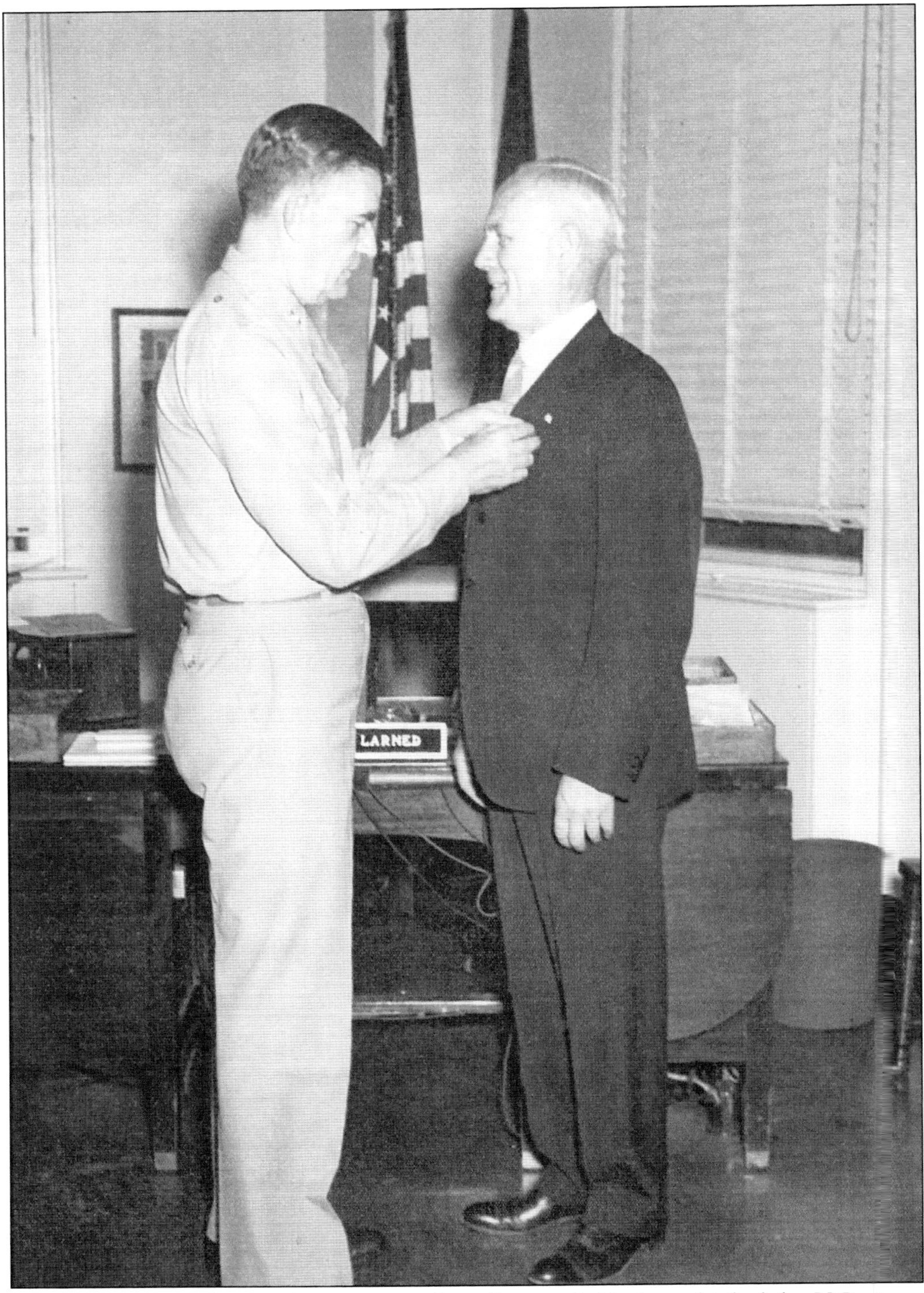

GEN. NORMAN F. RAMSEY DECORATED. Gen. Everett S. Hughes, chief of the U.S. Army Ordnance Department, decorates Gen. Norman F. Ramsey, retired, with the Legion of Merit during "General Hughes Day" in 1946. Ramsey was arsenal commander from 1922 to 1926. (U.S. Army Photograph-Picatinny Arsenal.)

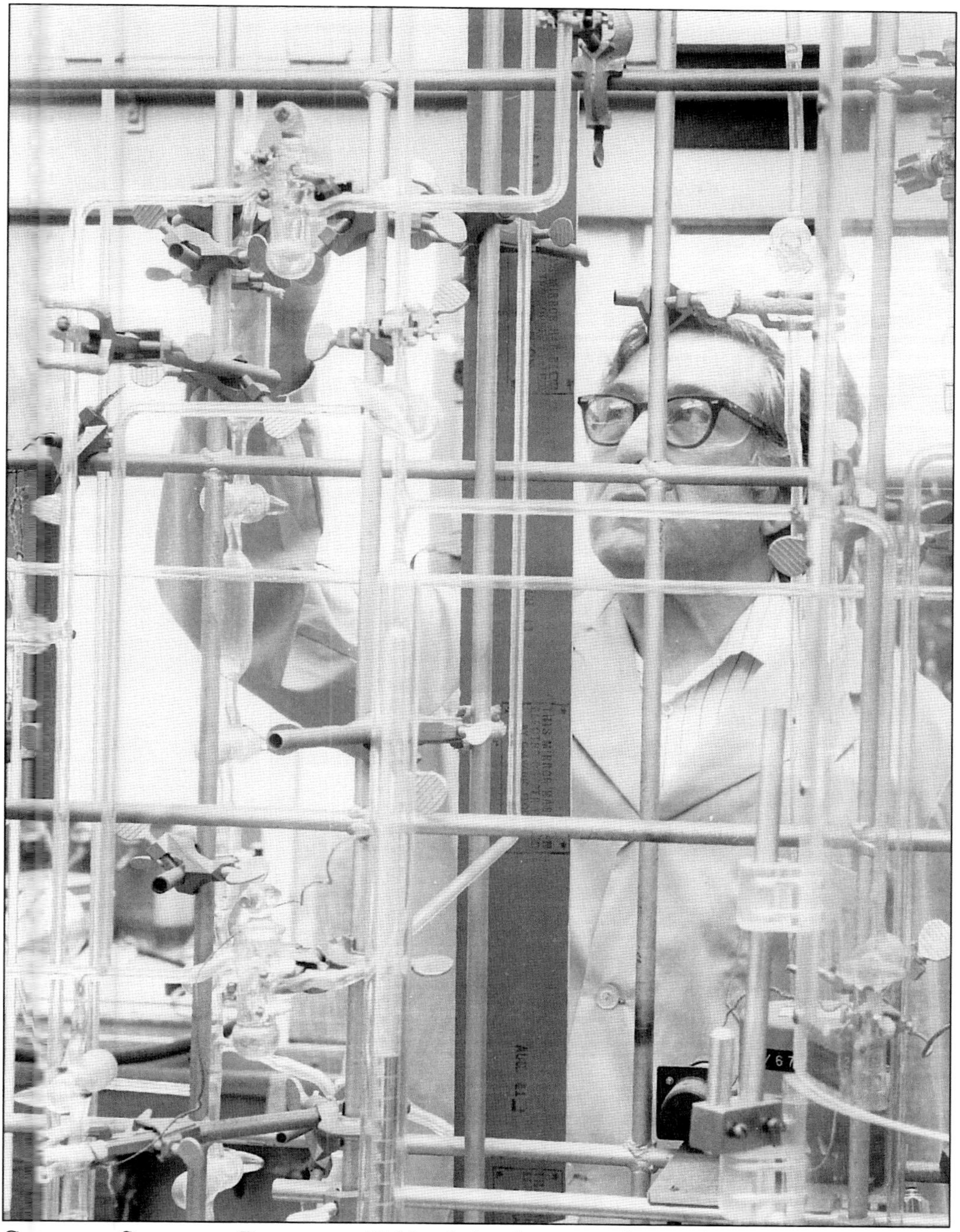

CHEMICAL SYNTHESIS RESEARCH. Hidden behind a maze of test tubes, a chemist works on chemical synthesis research seeking new molecules and a simple and more economical method for their formation. Sought was improved stability and behavior gas rather than higher energy content. (U.S. Army Photograph-Picatinny Arsenal.)

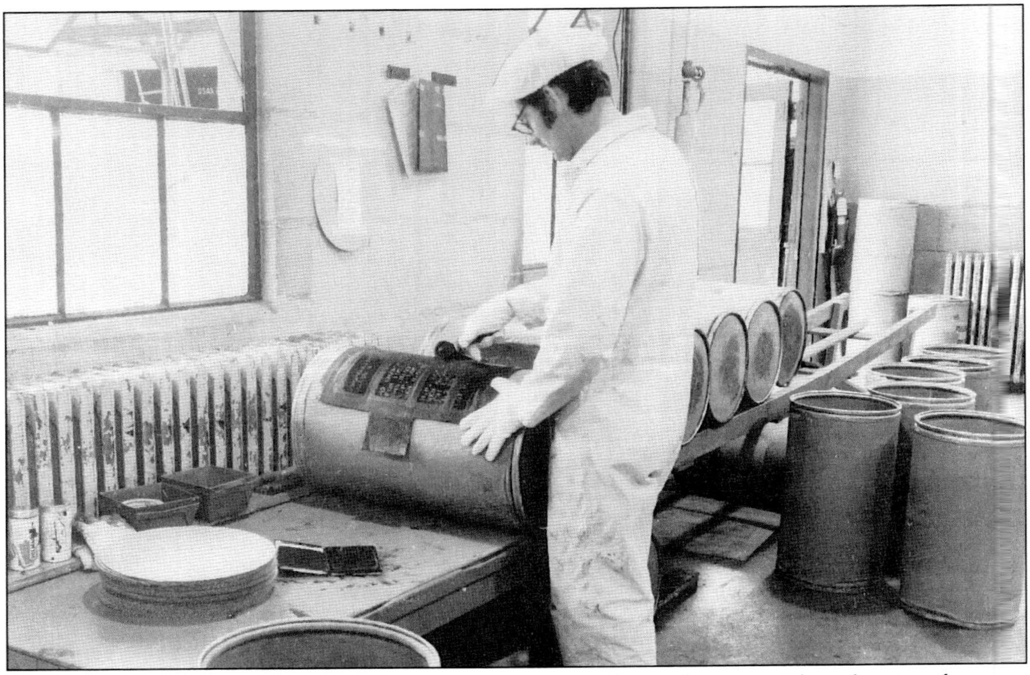

IDENTIFYING CONTAINERS. A workman uses a paint roller and pre-cut identification forms to stencil containers of 105-mm M67 charges in the arsenal's loading division in 1972, prior to shipment to the fighting front or storage. (U.S. Army Photograph-Picatinny Arsenal.)

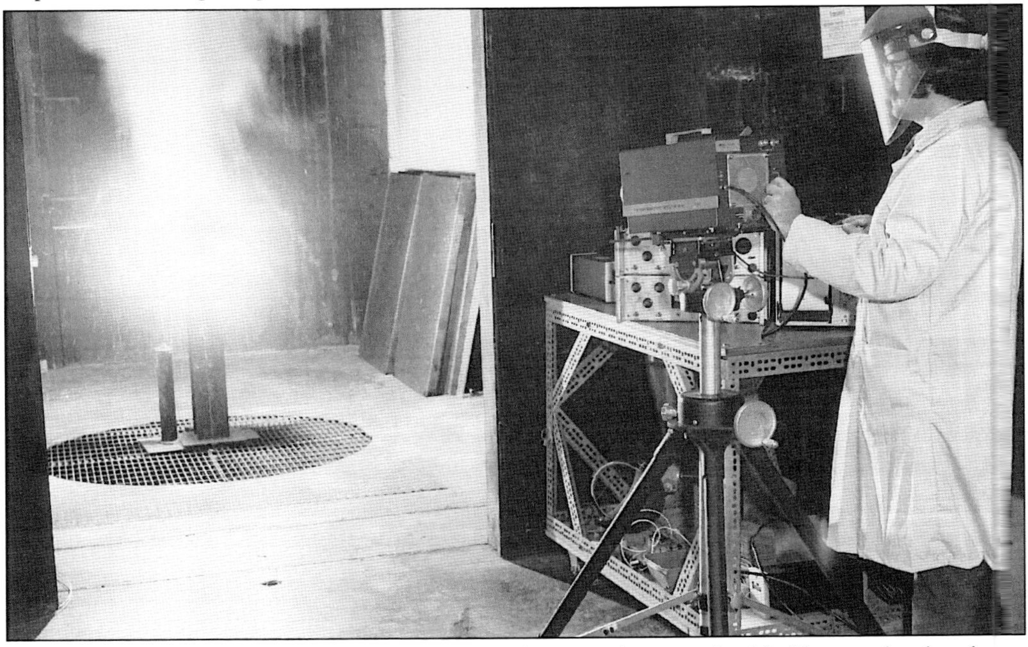

TESTING THE CANDLEPOWER OF FLARES. A technician, his eyes shielded by goggles, headgear, and heavy doors, tests an experimental flare to determine its intensity and spectral output. Pyrotechnic engineers produced two notable flares: one producing 800,000 candlepower during World War II, and a photoflash cartridge producing 20 million candlepower, visible 1,000 miles away for use in the space program. (U.S. Army Photograph-Picatinny Arsenal.)

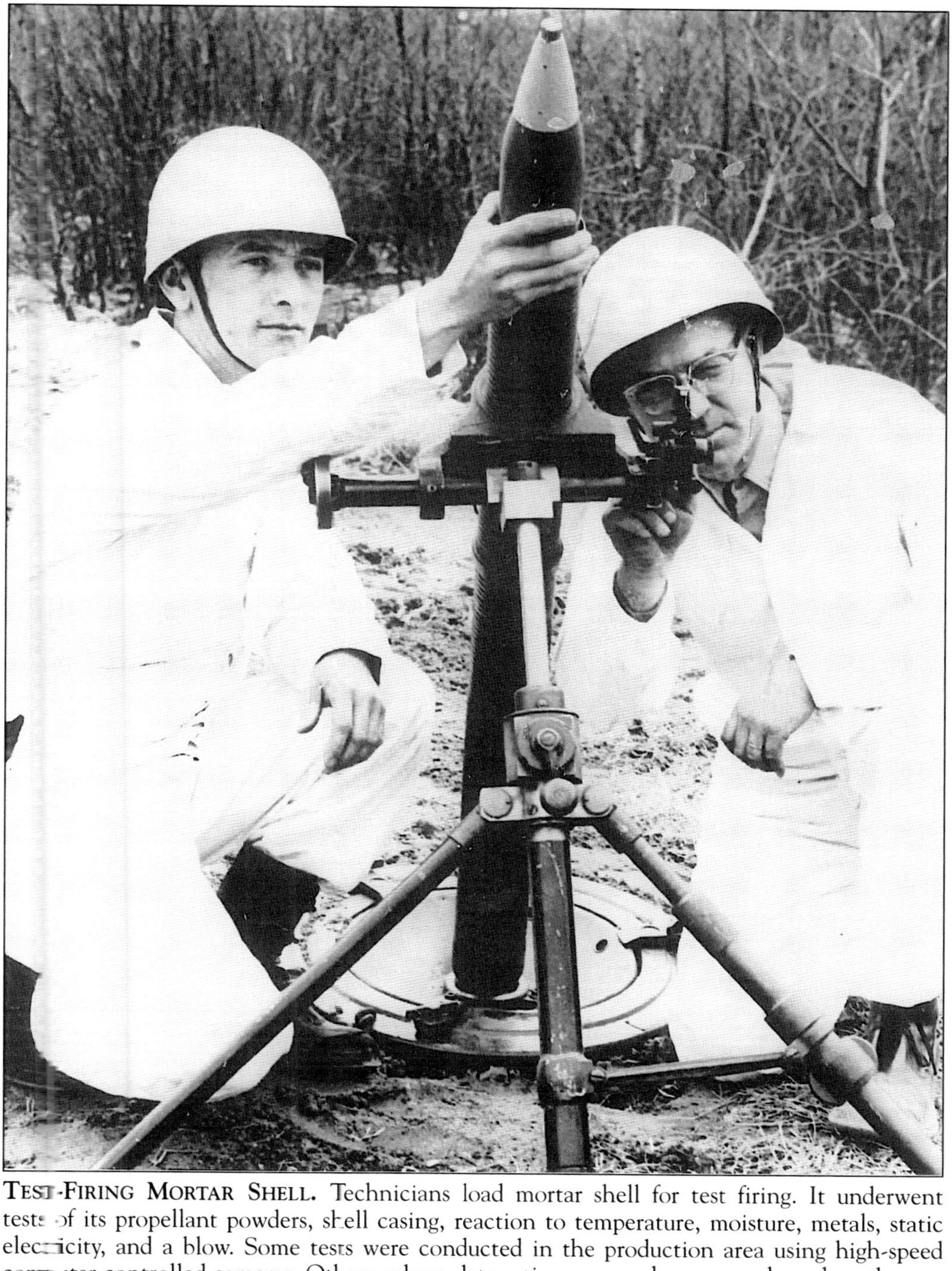

TEST-FIRING MORTAR SHELL. Technicians load mortar shell for test firing. It underwent tests of its propellant powders, shell casing, reaction to temperature, moisture, metals, static electricity, and a blow. Some tests were conducted in the production area using high-speed computer-controlled cameras. Others, where detonation occurred, were conducted on the test ranges. (U.S. Army Photograph-Picatinny Arsenal.)

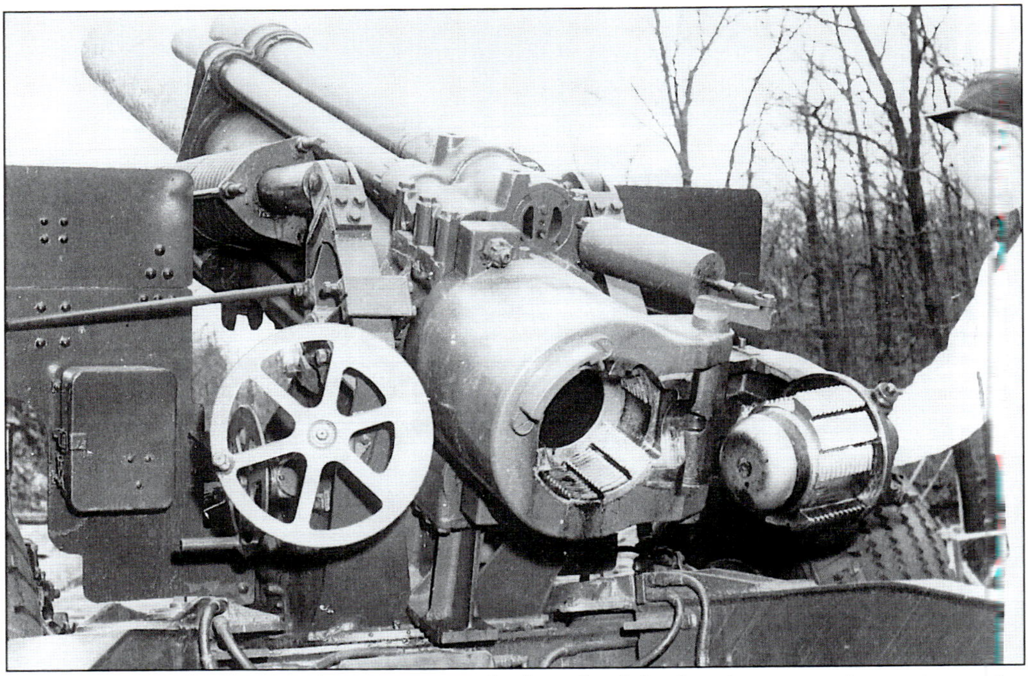

Test Range Firing. A technician opens the breech of the howitzer preparing to insert the artillery shell for test firing on one of the arsenal's test ranges. Smaller weapons, up to the .50-caliber machine gun, were fired, while high-speed cameras captured the action of the bullet in flight to determine aerodynamic co-efficient, flow characteristics, causes of dispersions, and yaw. (U.S. Army Photograph-Picatinny Arsenal.)

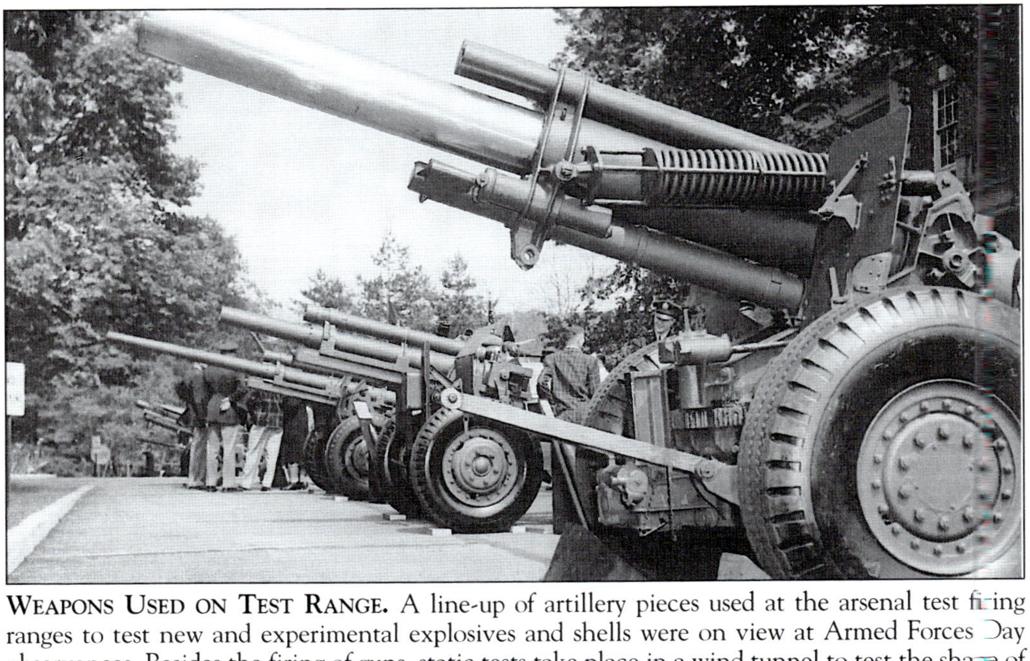

Weapons Used on Test Range. A line-up of artillery pieces used at the arsenal test firing ranges to test new and experimental explosives and shells were on view at Armed Forces Day observances. Besides the firing of guns, static tests take place in a wind tunnel to test the shape of the projectile or bomb in relation to wind pressure. (U.S. Army Photograph-Picatinny Arsenal.)

LOADING MORTAR SHELL AT TEST RANGE. A worker loads an experimental mortar shell for test firing at the test range. The mortar is enclosed in a reinforced cement bunker with observation ports at the rear. Overhead arc lights provide illumination for night testing of ammunition. (U.S. Army Photograph-Picatinny Arsenal.)

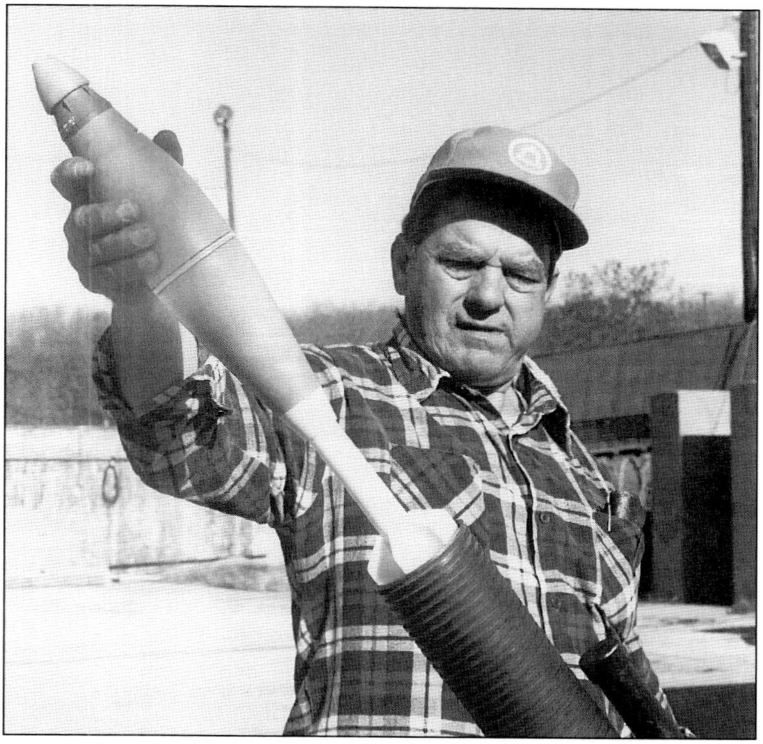

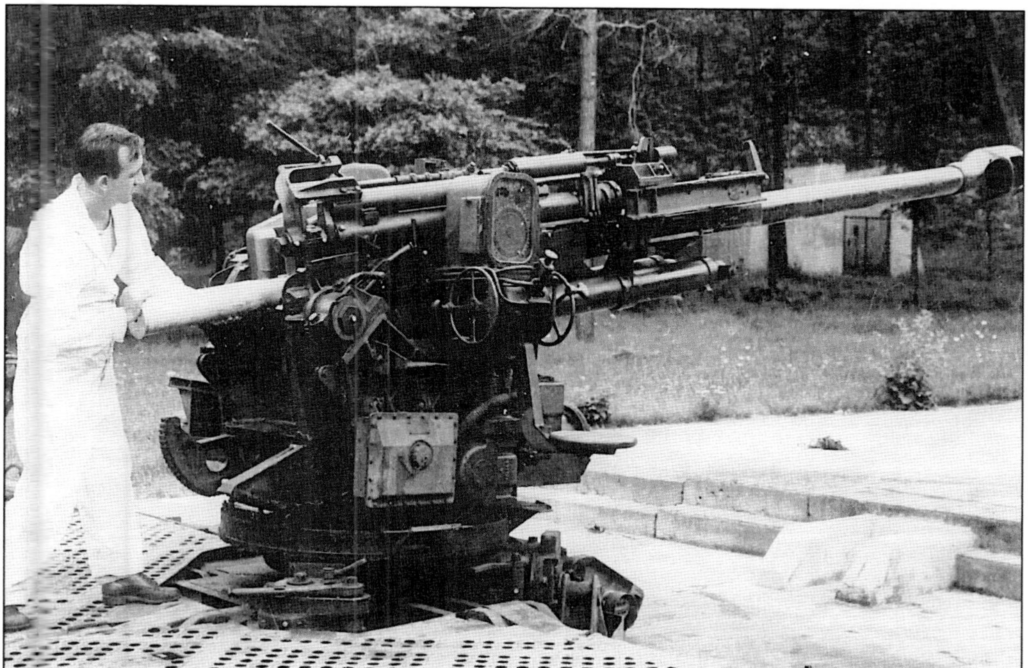

TESTING ANTI-AIRCRAFT GUN AMMUNITION. A technician loads an experimental shell into a 90-mm anti-aircraft gun mounted on a babbette at the test firing range. Tests are conducted on both experimental munitions and selected ammunition taken from production lines and storage to insure quality. (U.S. Army Photograph-Picatinny Arsenal.)

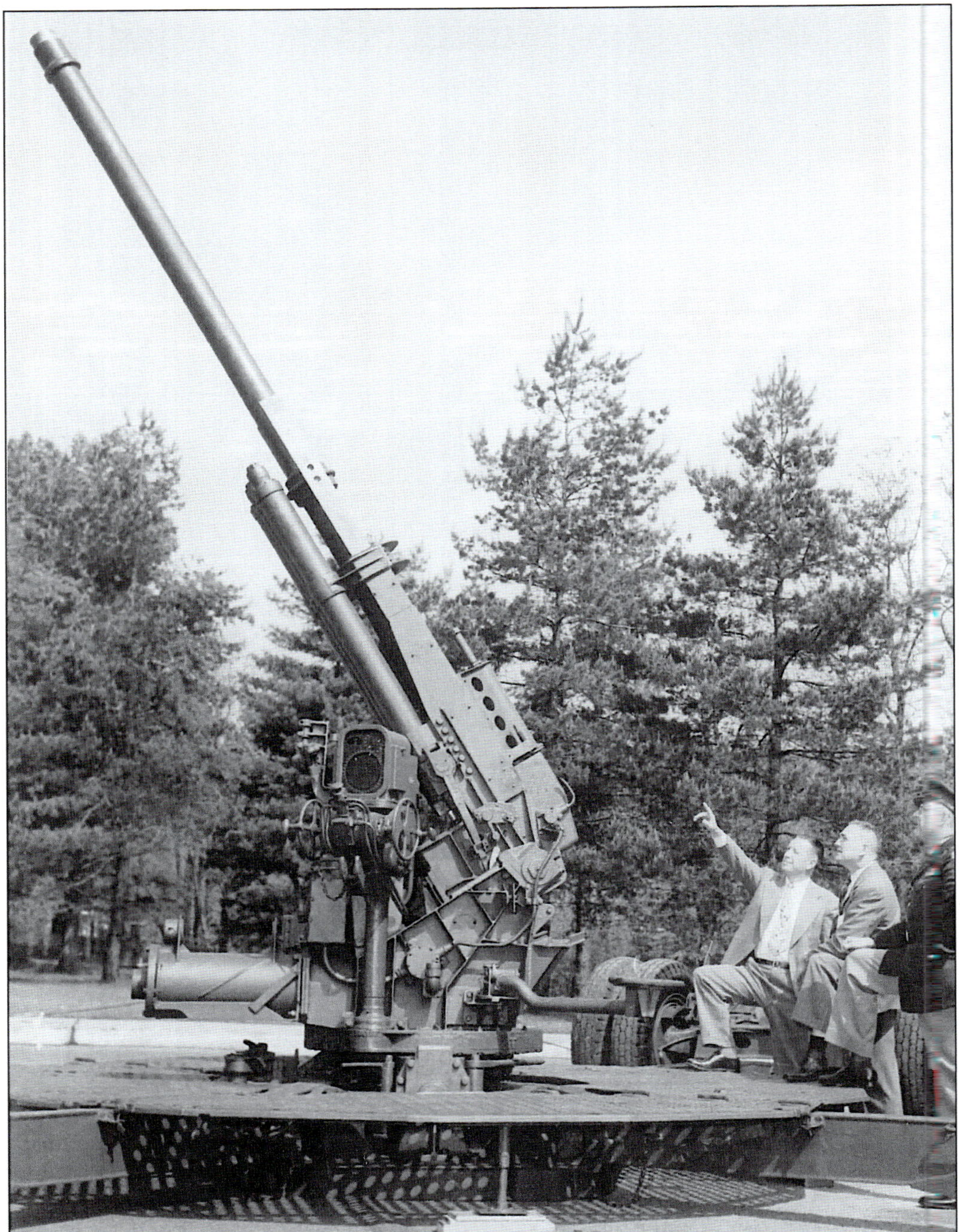

EXAMINING AN ANTI-AIRCRAFT GUN. An army officer and two civilians examine a mobile 90-mm anti-aircraft gun mounted on a babbette at the arsenal. Improvement of ammunition for this weapon is one of the arsenal's missions of research, design, and development. (U.S. Army Photograph-Picatinny Arsenal.)

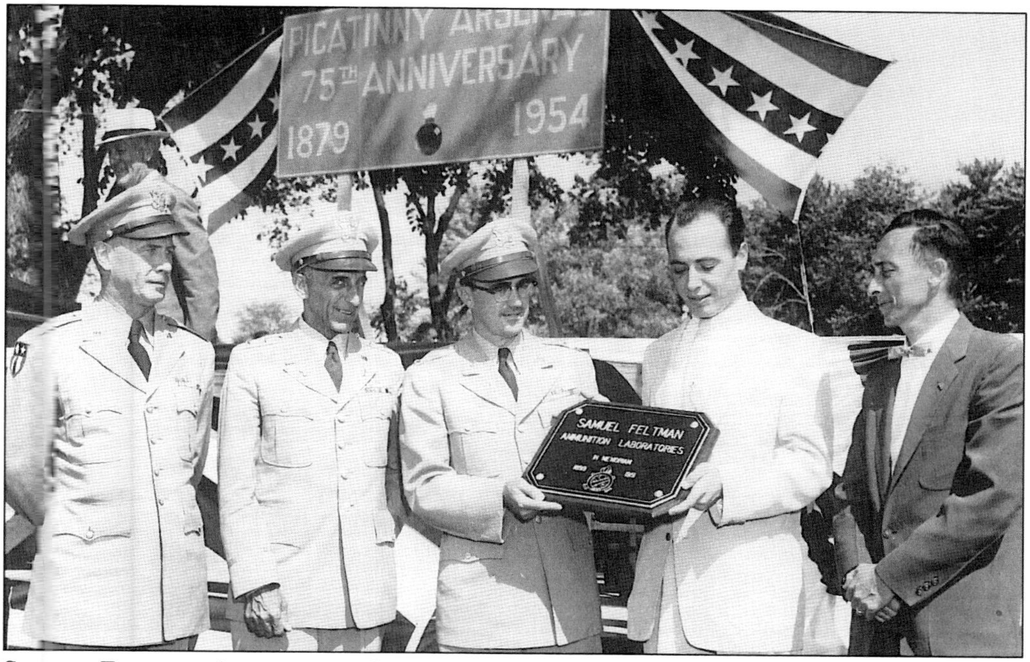

SAMUEL FELTMAN AMMUNITION LABORATORIES. Named for Samuel F. Feltman, a pioneer in the development of ballistic missiles, this building was dedicated during the 75th anniversary observance of the arsenal in 1954. Col. John D. Armitage is presenting a plaque to David F. Feltman, son of Samuel F. Feltman, while his brother, Dr. Robert F. Feltman, observes. (U.S. Army Photograph-Picatinny Arsenal.)

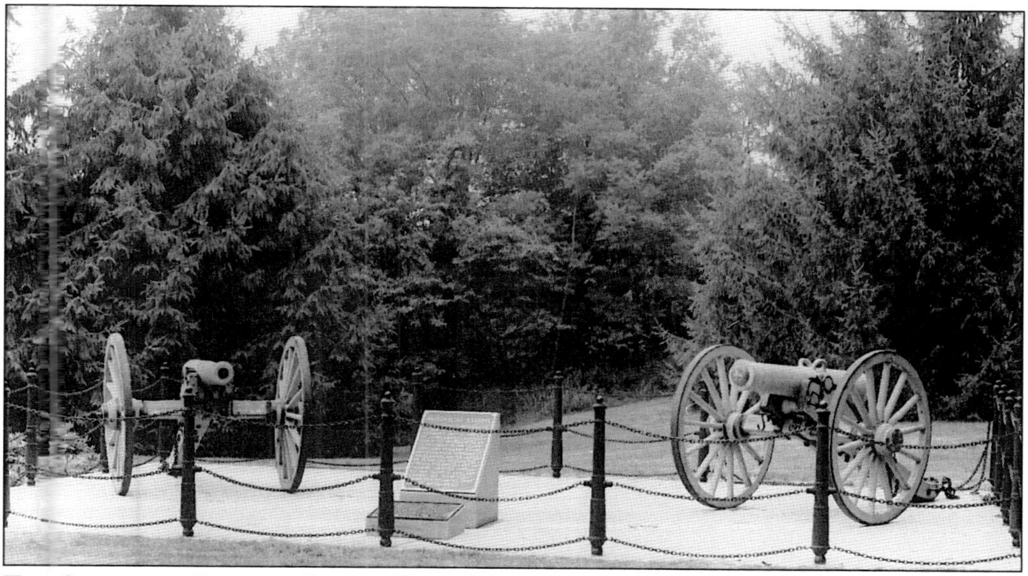

THE SARATOGA CANNON PARK. This exhibit at the arsenal's main entrance was originally part of the Saratoga Park at Frankford Arsenal, Philadelphia. It was moved to Picatinny in 197[illegible], shortly after establishment of the U.S. Army Armament Research & Development Command here and the closure of Frankford Arsenal. It commemorates the surrender of Gen. John Burgoyne's 7,000-man British Army in the Revolutionary War. (U.S. Army Photograph-Picatinny Arsenal.)

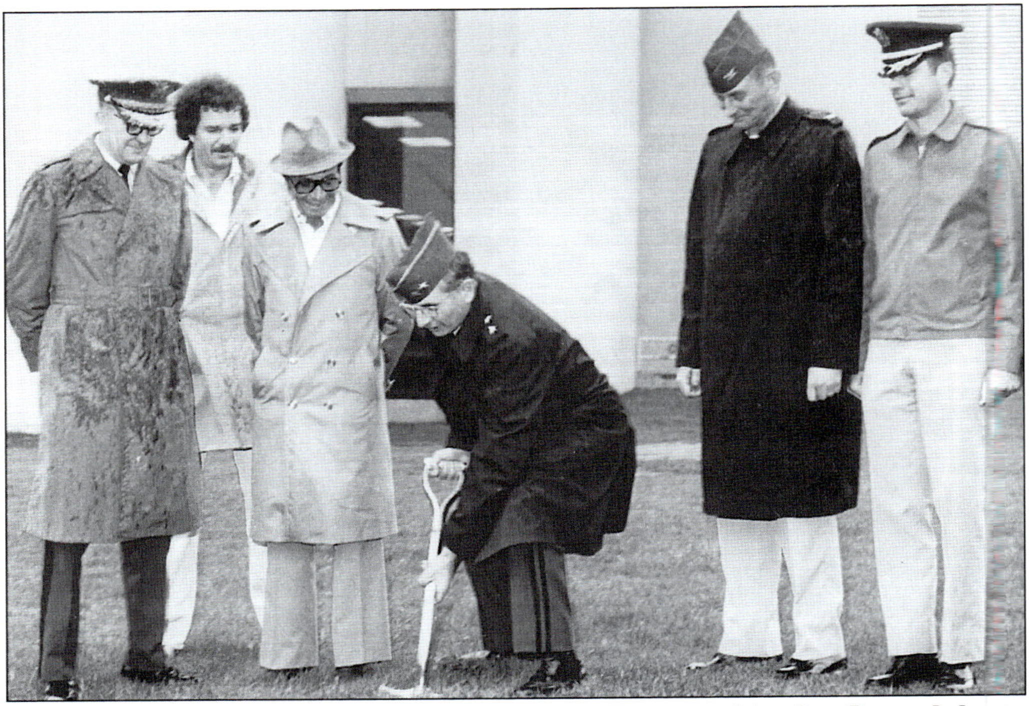

BREAK GROUND FOR ARRADCOM ADMINISTRATION BUILDING. Maj. Gen. Bennet L. Lewis, first commander of U.S. Research & Development Center at arsenal, breaks ground for a new four-story $7.9 million administration and engineering building, June 18, 1979. It is named for Maj. Francis H. Parker, the first commanding officer of Picatinny, who was responsible for the acquisition of the land for the arsenal in 1880. (U.S. Army Photograph-Picatinny Arsenal.)

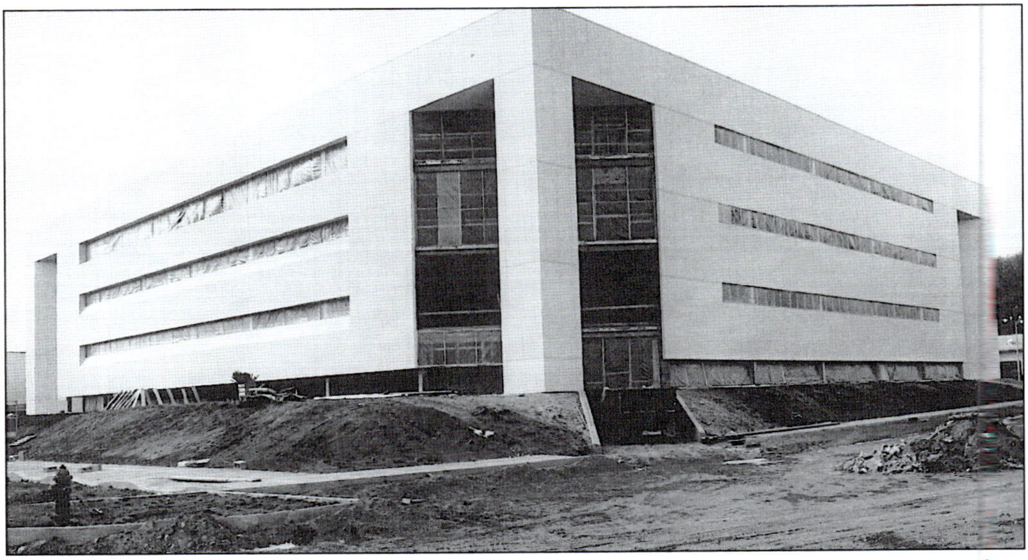

UNDER CONSTRUCTION. This photograph shows the new administration center with a 930-person capacity under construction. The structure, which superceded Building 151, the administration center from 1929 to 1981, is joined to building 94 by a conference center, seating 450 people. (U.S. Army Photograph-Picatinny Arsenal.)

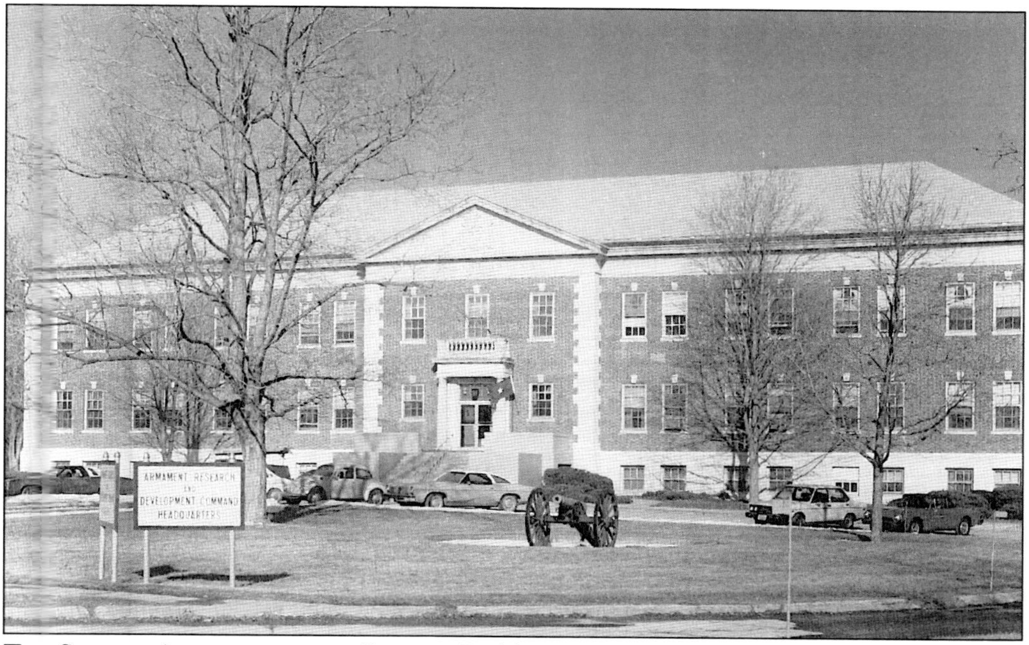

THE SECOND ADMINISTRATION CENTER. Building 151, a two-story brick building constructed in 1929-30, now serves as the Armament Research & Development Command headquarters. The insignia engraved on the pediment over the main entrance is the seal of the U.S. Army Ordnance Corps. See page 34 for the ruins of the first administration center, damaged by the 1926 explosion. (U.S. Army Photograph-Picatinny Arsenal.)

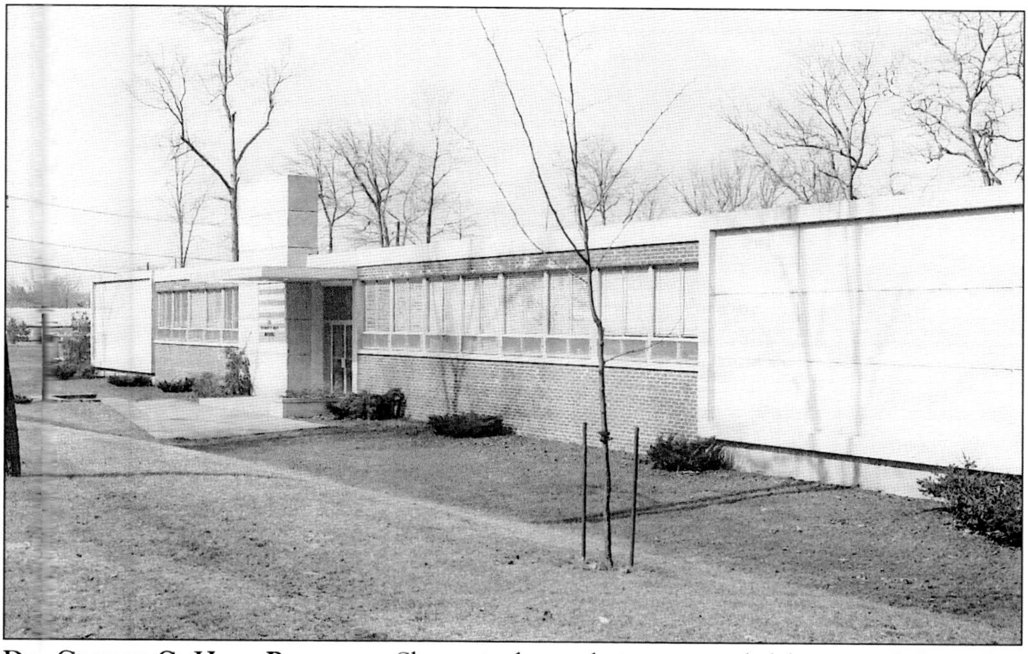

DR. GEORGE C. HALE BUILDING. Shown is the explosives research laboratory dedicated in May 1962, to Dr. Hale, the arsenal's chief chemist, who spent 30 years at Picatinny. It was described at that time as the finest explosive research laboratory in the world. (U.S. Army Photograph-Picatinny Arsenal.)

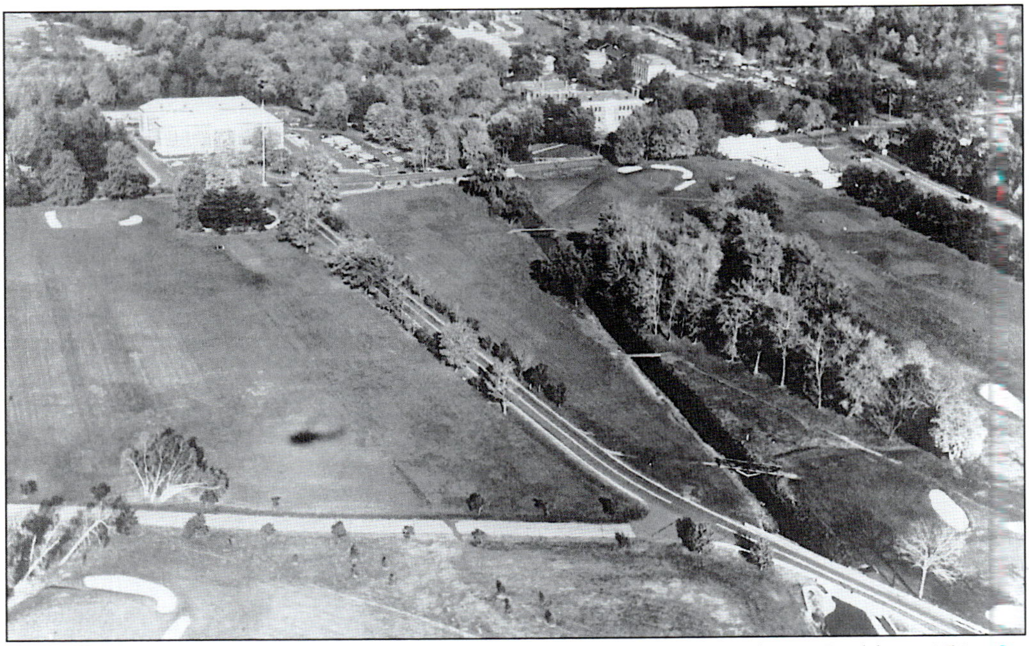

APPROACH TO BUILDING 151. This aerial view of the approach to Building 151, the administrative center of the arsenal (1929–1981), shows the rear addition, which was made at the start of World War II. Green Pond Brook is beside the main entrance road. On either side of the road are the fairways, greens, and sand traps of the golf course. (U.S. Army Photograph-Picatinny Arsenal.)

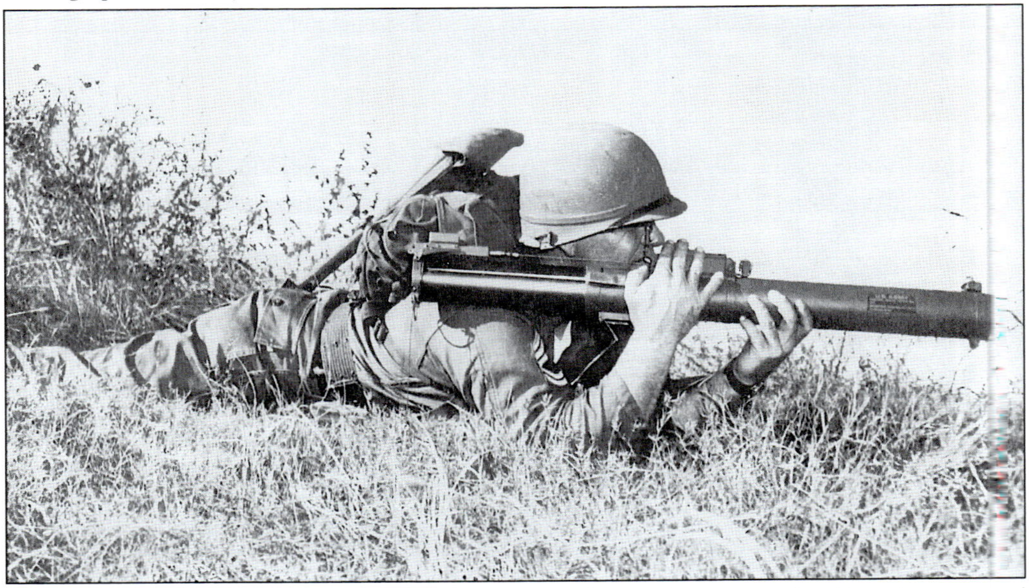

LAW LIGHT ANTI-TANK WEAPON. This weapon, better known as the Bazooka, stopped Rommel's Tank Corps in Africa. It failed, however, to halt Russian T-34 tanks used by North Korea. Rocket shells for a heavier 3.5-inch Bazooka were being tested on a pilot line at the arsenal. Overnight it was geared to full production. Finished rockets were flown from the Morristown, NJ airport halfway around the world to Korea. Within seven days, Red Tanks were destroyed. (U.S. Army Photograph-Picatinny Arsenal.)

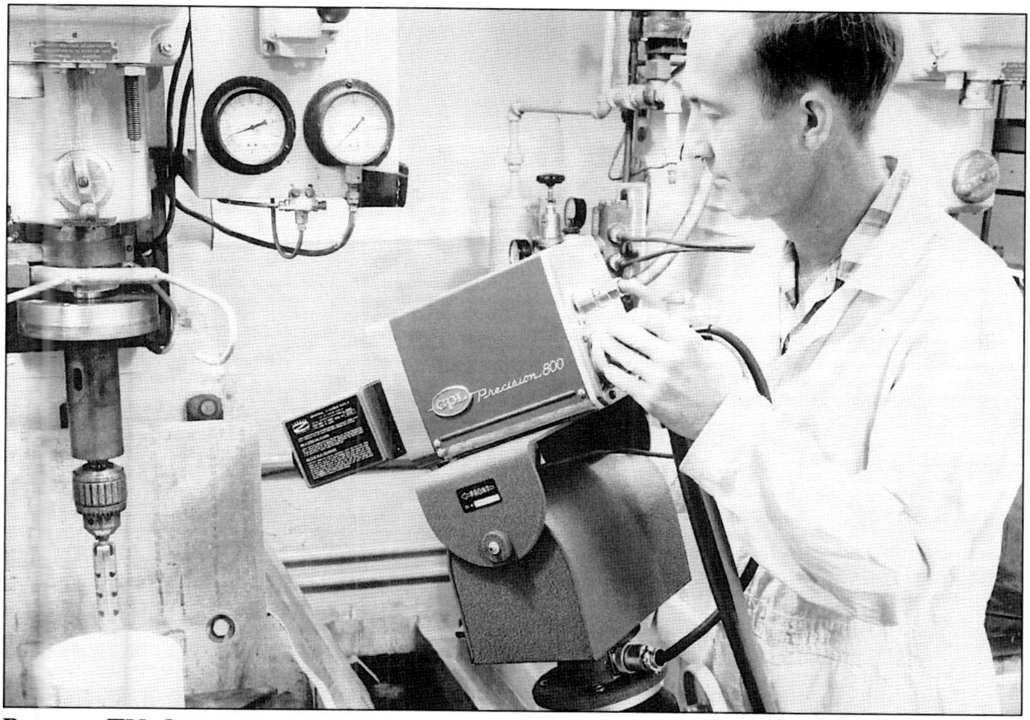

REMOTE TV OBSERVATION. Dangerous operations, especially those involving experimental nuclear munitions, are observed by technicians and scientists through remote television cameras, which record both the test and the read-out on gauges recording it. (U.S. Army Photograph-Picatinny Arsenal.)

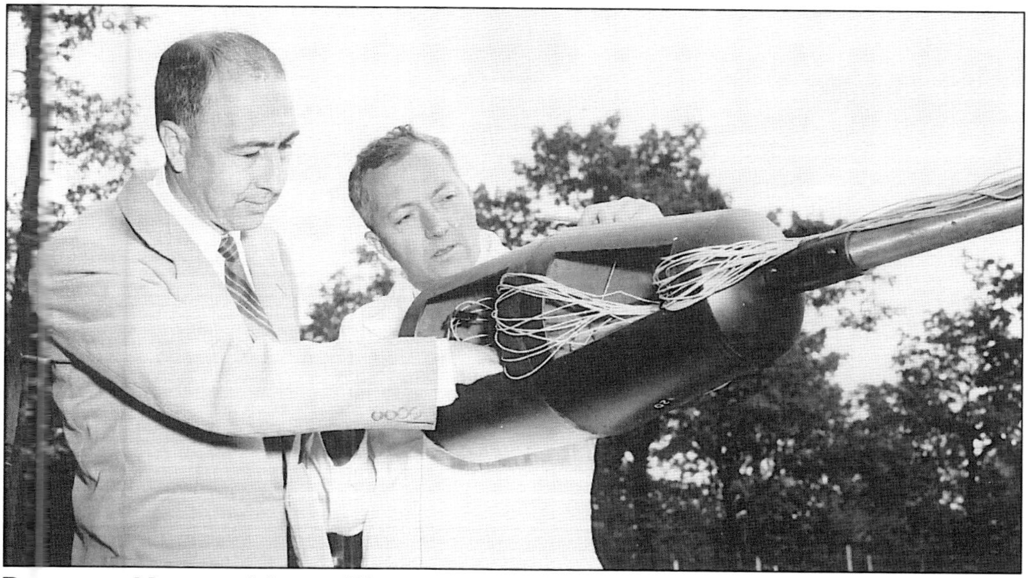

PREPARING NUCLEAR MISSILE WARHEAD. A Technical Service Directorate engineer prepares a nuclear missile warhead section for testing in the radio frequency laboratory. Missile systems for which the arsenal designed the "business end" of the system ranged from the Nike-Ajax, the country's first supersonic air defense guided missile, to the Nike-Hercules and the Nike-Zeus. (U.S. Army Photograph-Picatinny Arsenal.)

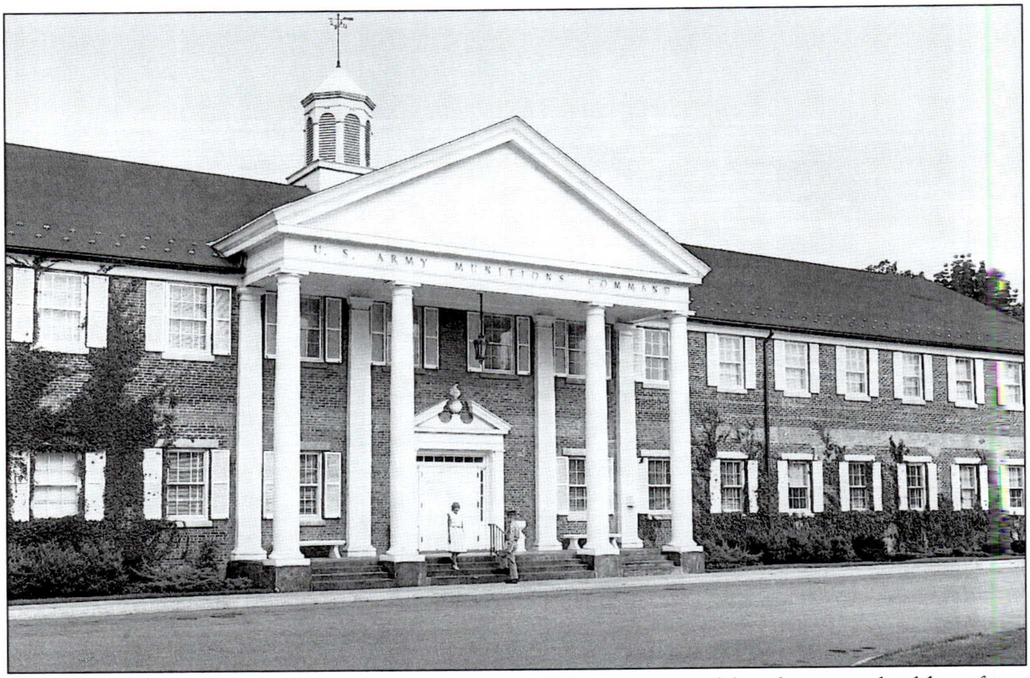

U.S. MUNITIONS COMMAND. This was the munitions command headquarters building from 1962 to 1973. Built in 1943, it is now headquarters for ground combat and support systems, the program executive office, program managers for arms, and the Crusader and Paladin weapons systems. (U.S. Army Photograph-Picatinny Arsenal.)

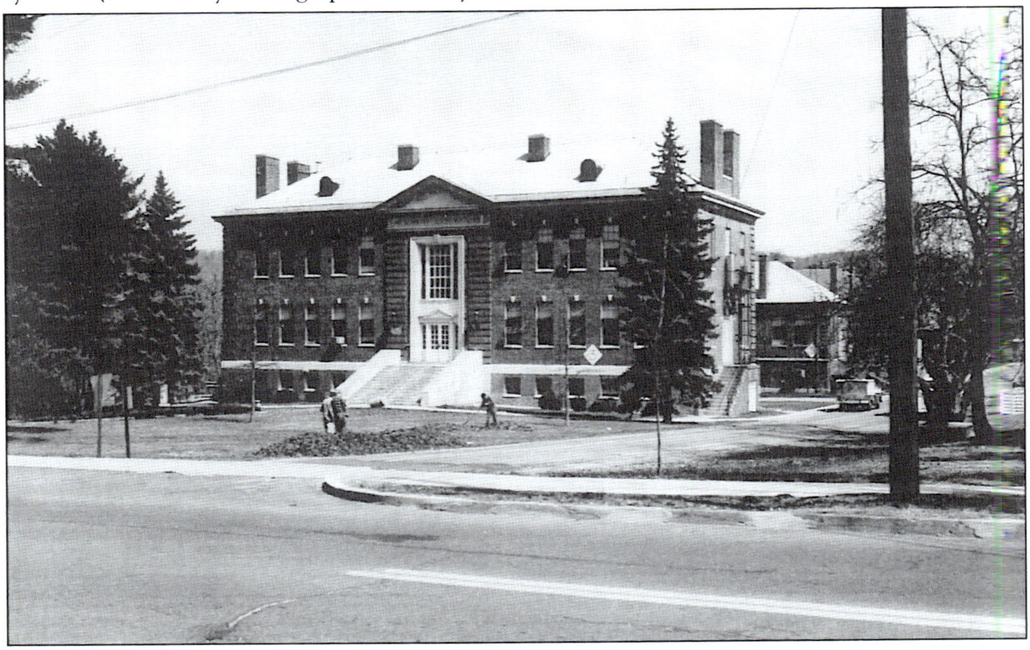

CHEMISTRY LABORATORY. Adjacent to Building 151, the arsenal's administrative center from 1929 to 1981, the 1930 brick two-story building was expanded with an addition to the rear in 1942. Much of the development of World War II weapons was conducted in its laboratories. (U.S. Army Photograph-Picatinny Arsenal.)

DRAFTING. This is where it all begins—putting ideas for new ammunition and missiles on paper before test models are made, tested, and evaluated. After the 1926 explosion, the brick magazines, which survived the blast, were re-roofed and re-windowed, and in them a different type of drafting was done: plans for the administration building and chemical laboratory group, the nucleus of the new Picatinny. (U.S. Army Photograph-Picatinny Arsenal.)

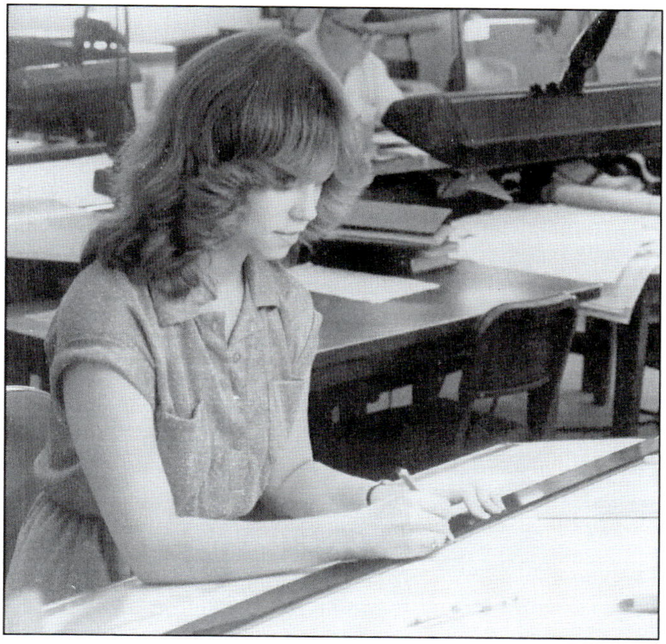

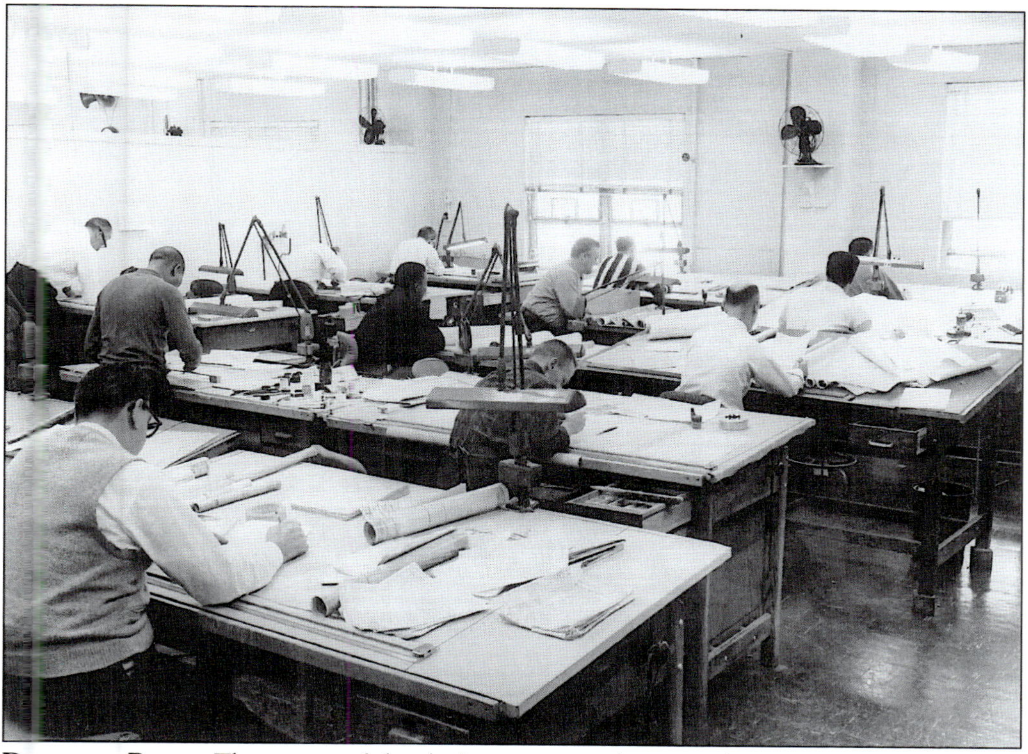

DRAFTING ROOM. The corner of the drafting room at the arsenal in 1965, when plans were shifting from conventional munitions to nuclear weapons, was where, in 1952, plans were drafted for one of the arsenal's most notable achievements—the development of the first nuclear shell capable of being fired from an artillery gun. (U.S. Army Photograph-Picatinny Arsenal.)

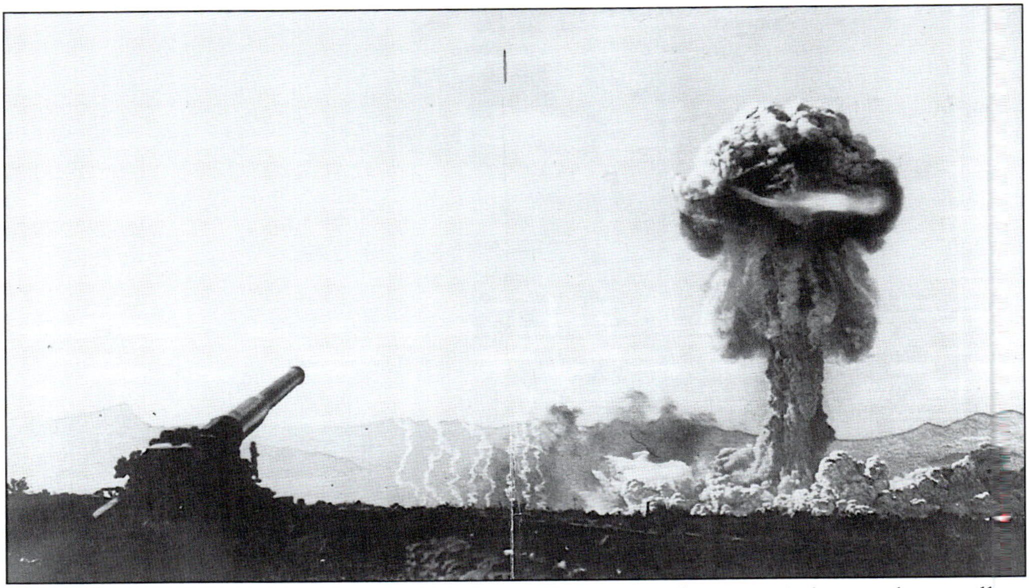

TEST FIRING THE FIRST NUCLEAR SHELL. "Atomic Annie," the nation's first nuclear artillery shell, is being test fired at the Las Vegas, Nevada, proving grounds. To meet the need for a smaller shell, development of an 8-inch nuclear shell capable of being used by both the army and navy was started. Annie was the first shell to use the lightweight, high strength metal titanium. (U.S. Army Photograph-Picatinny Arsenal.)

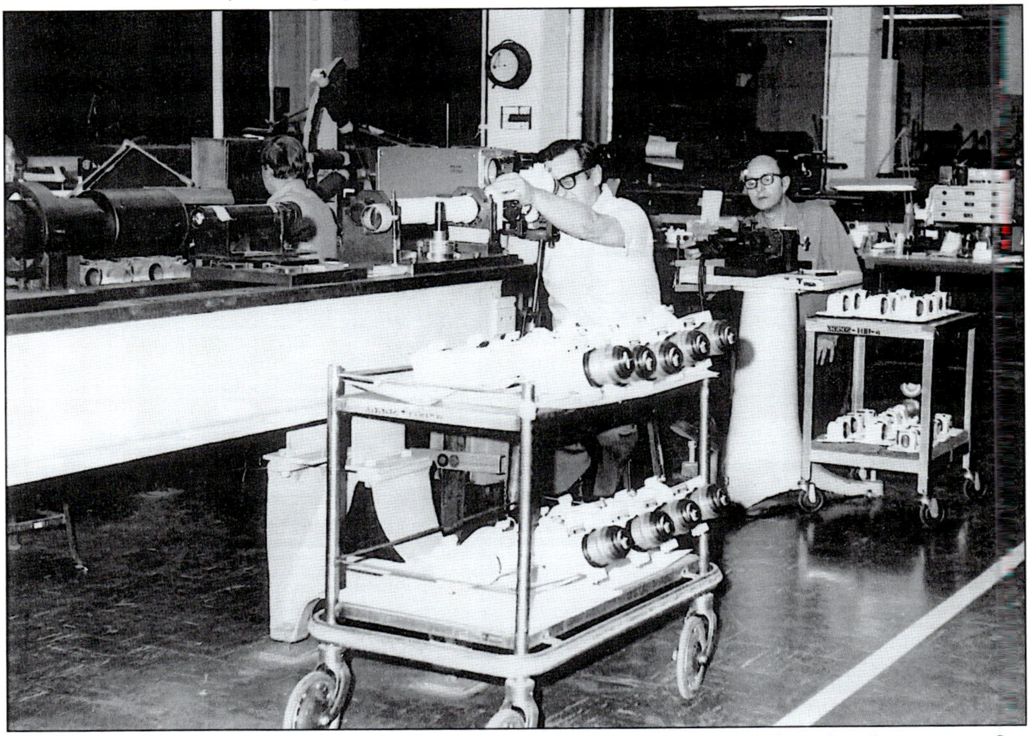

THE TELESCOPE SHOP. Lens, used in regular aiming devices for a multitude of weapons, fire control devices, close-up examination of test results, and use in new weapons, is centered in the telescope shop in Building 95. (U.S. Army Photograph-Picatinny Arsenal.)

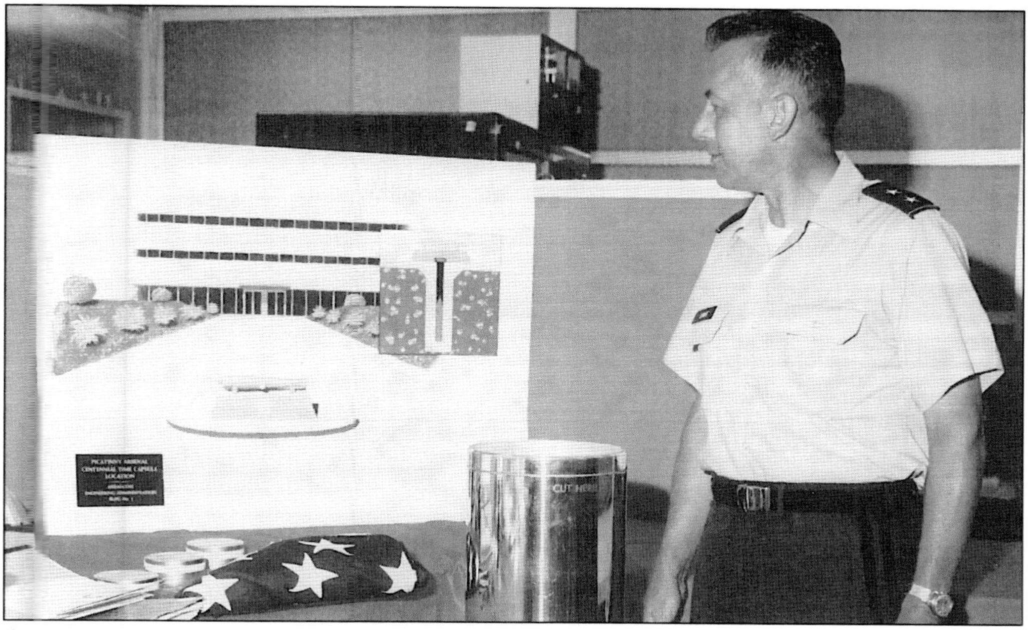

CENTENNIAL TIME CAPSULE. Gen. Allen H. Light Jr., ARRADCOM commander, looks at an artist's rendering of where Picatinny Centennial Time Capsule was to be buried in front of the new administration center. The capsule was buried, but the centerpiece, featuring a Civil War- era cannon, was never constructed. (U.S. Army Photograph-Picatinny Arsenal.)

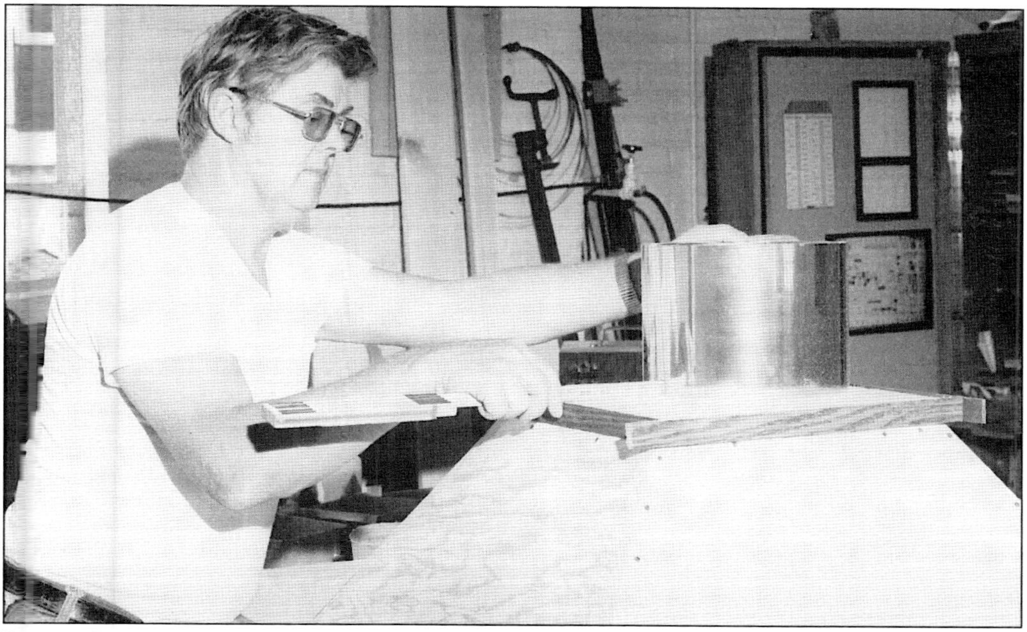

IRVING STANFORD, EXHIBIT SPECIALIST, ASSEMBLES A TIME CAPSULE. The 718 individual items in the capsule consisted of items of general and historical interest, such as photos, biographies, reports, technical manuals, descriptions of munitions developed at the arsenal, microfilm of operations, booklets, the American flag flown in 1980 at the Research and Development Command, guides to the arsenal, Picatinny newspapers, and selected ammunition. (U.S. Army Photograph-Picatinny Arsenal.)

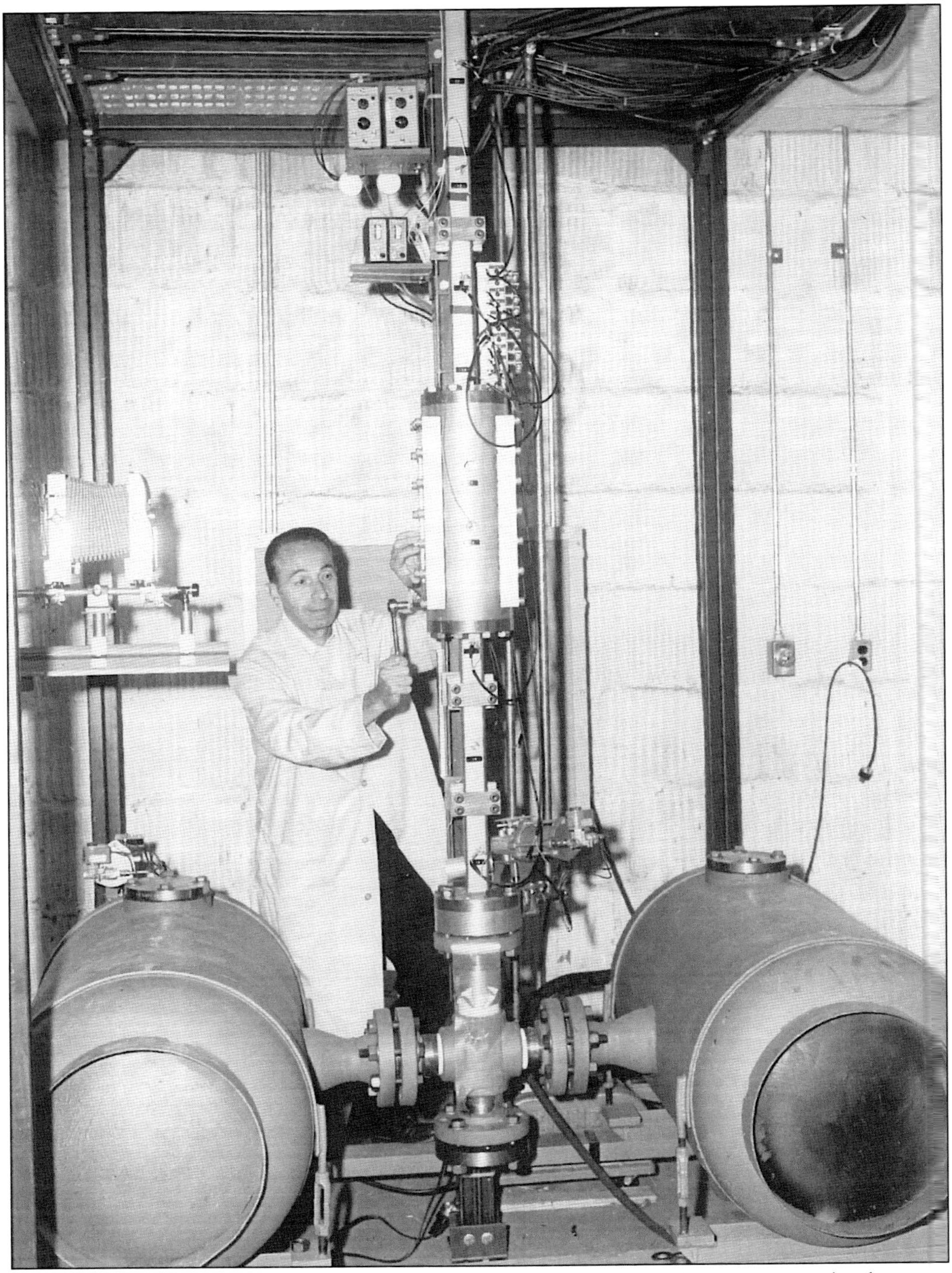

A Test for Shock Waves. A technician adjusts a particle shock tube to generate shock waves, simulating those created by the detonation of conventional ammunition. This unit, together with sound tunnels, radio frequency waves, TV and high-speed cameras that photograph munitions in flight, and spectrometers, is typical of modern devices used in the development of conventional and nuclear munitions. (U.S. Army Photograph-Picatinny Arsenal.)

Six

The Public Gets a Peek

Armed Forces Day and the Museum

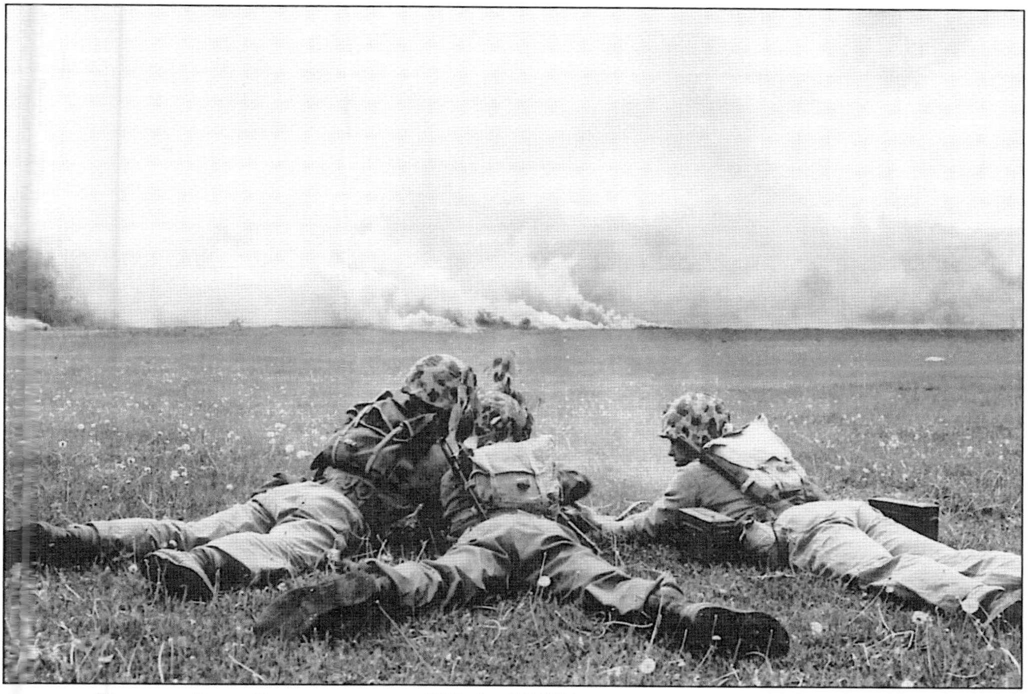

ARMED FORCES DAY OBSERVANCES. While 10,000 spectators watched, a team of Marine riflemen dropped by helicopter fire toward enemy soldiers, who were partially concealed by smoke screen in the annual observance of Armed Forces Day, held at the arsenal from 1950 to 1995. (U.S. Army Photograph-Picatinny Arsenal.)

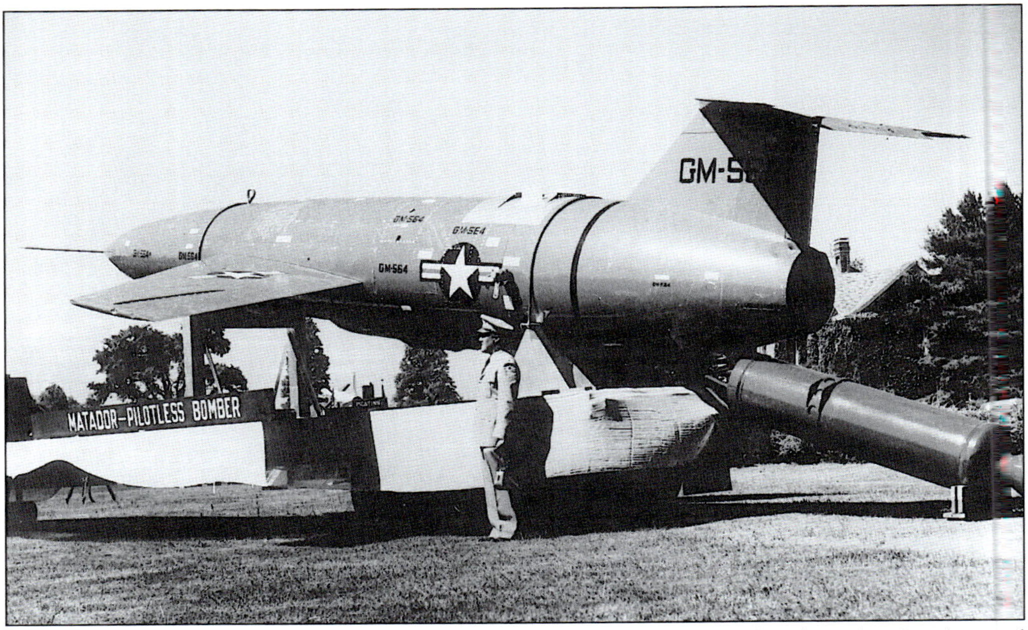

A PILOT-LESS BOMBER. This photograph shows a Matador pilot-less bomb with a Jet Assisted Take-Off booster on exhibit at the 75th anniversary of the founding of the arsenal July 2, 1954. Picatinny scientists and technicians helped develop the JATO engine. (U.S. Army Photograph-Picatinny Arsenal.)

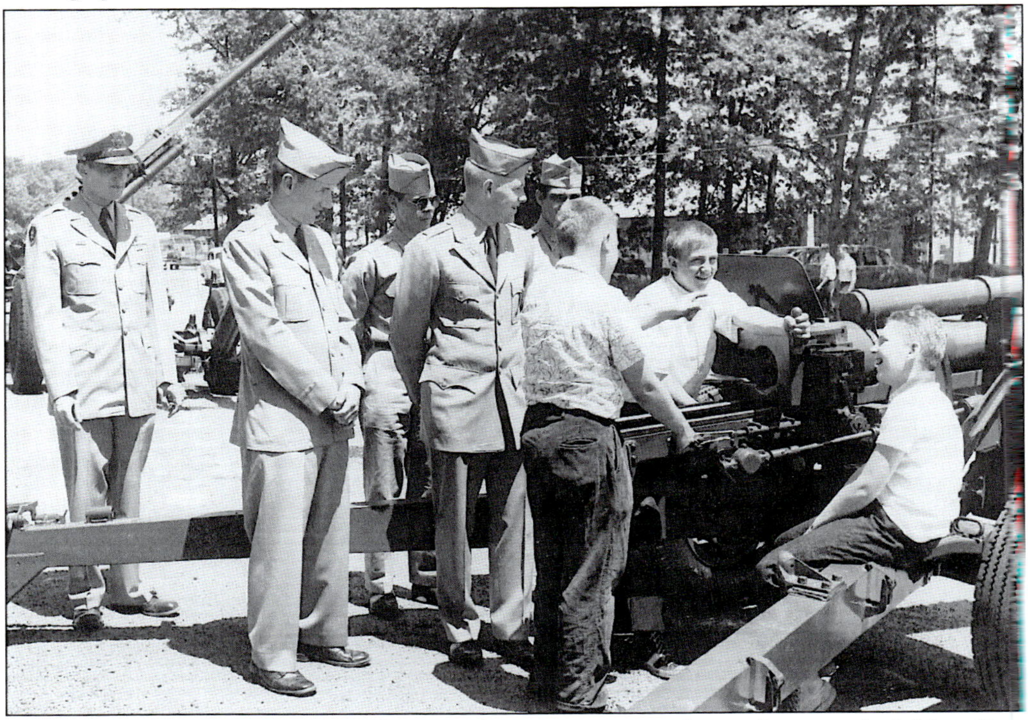

EXPLANATION. Army Ordnance officers explain to youngsters the mechanism of a field artillery gun on exhibit at Armed Forces Day observances at the arsenal in 1953. In the background is an anti-aircraft gun. (U.S. Army Photograph-Picatinny Arsenal.)

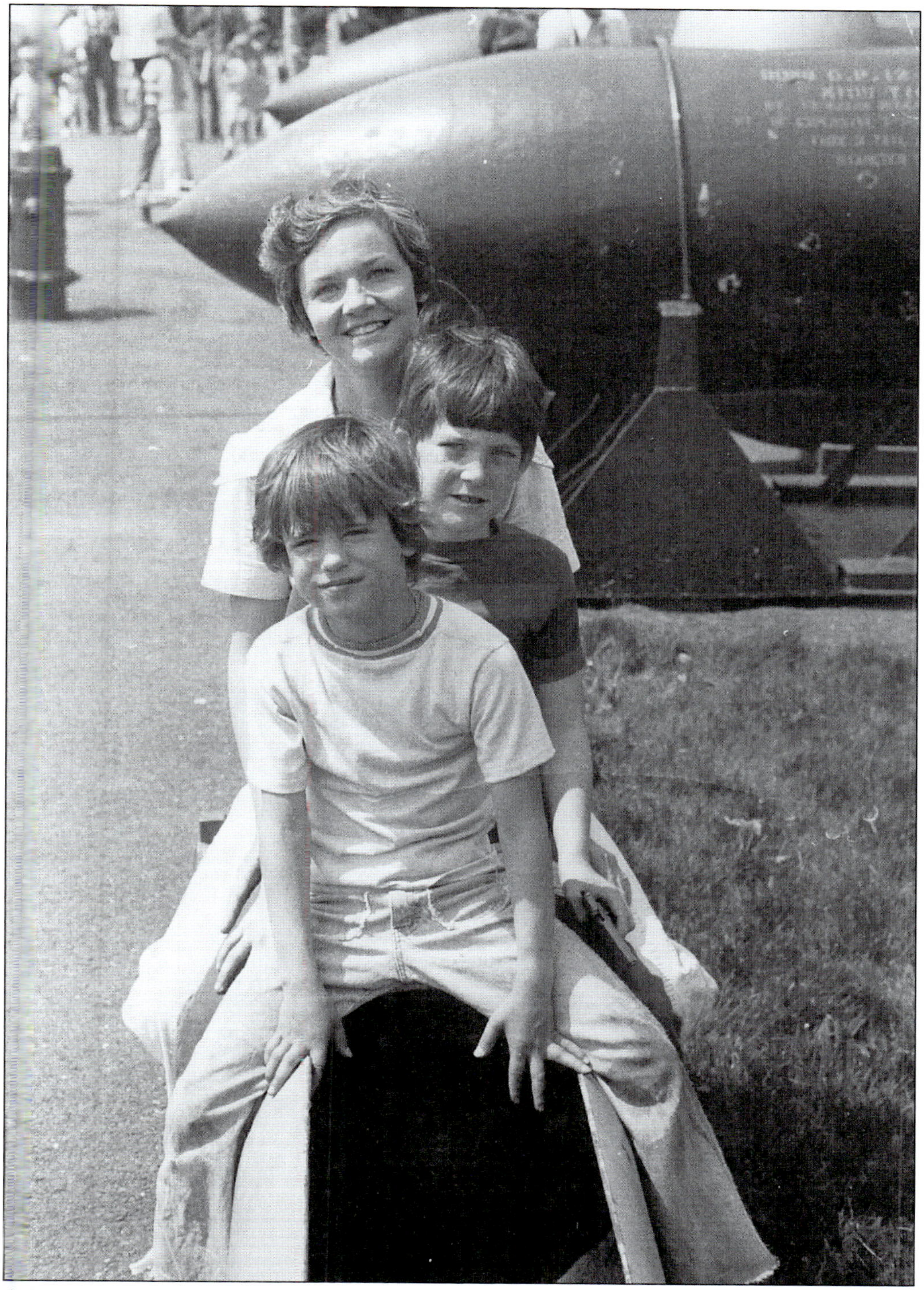

A JAPANESE TORPEDO. Mrs. Adelaide Cope, Chris Cope, and Ted Hall sit astride a Japanese torpedo in the arsenal museum park during Armed Forces Day observances on May 20, 1978. In the background are large World War II bombs, many of which were manufactured and filled with explosives manufactured at the arsenal. (U.S. Army-Picatinny Arsenal.)

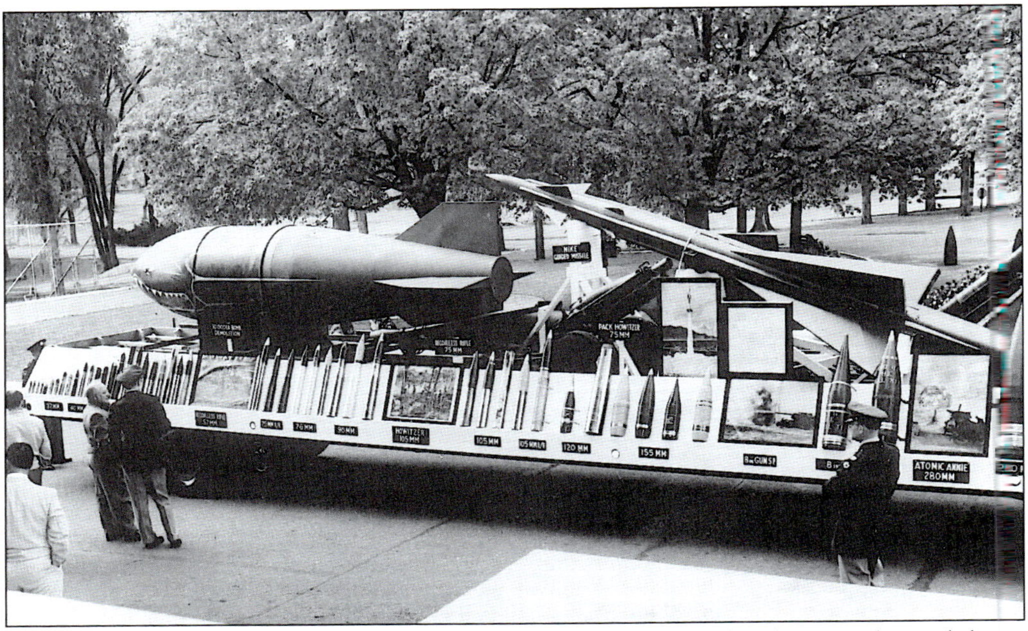

A FLOAT EXHIBIT. This float displayed rockets and shells developed at the arsenal on exhibit at Armed Forces Day observances in 1954. Shown are the Nike guided missile, "Atomic Annie," the 280-mm artillery shell developed to contain a nuclear warhead, recoilless rifle ammunition, and a multitude of shells. (U.S. Army Photograph-Picatinny Arsenal.)

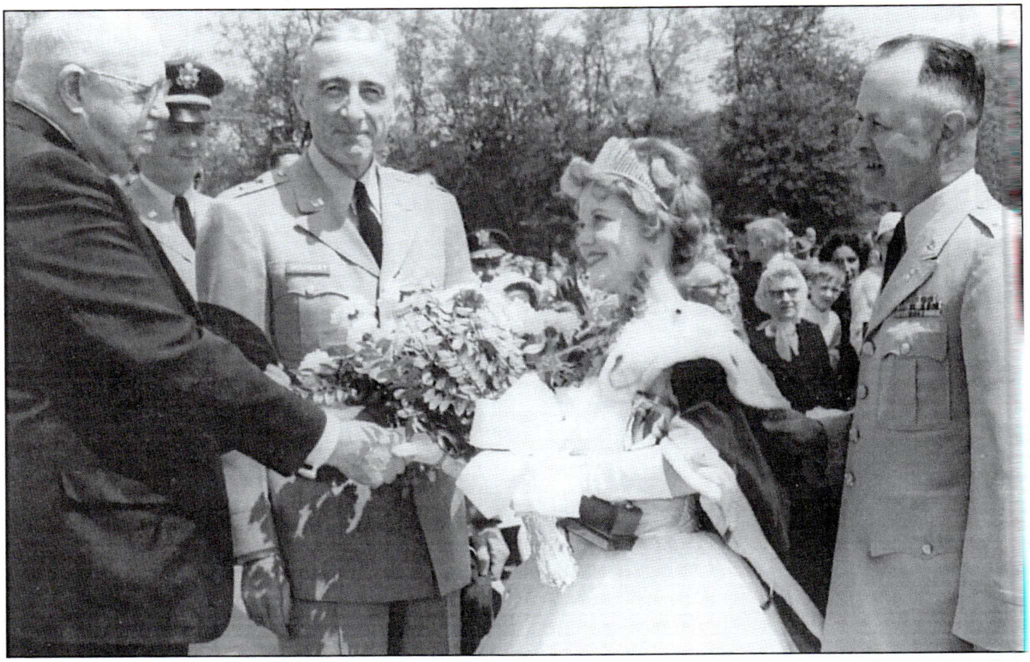

MISS PICATINNY ARSENAL. Dover, New Jersey mayor John Roach Jr. congratulates Miss Pat Vanorald, crowned "Miss Picatinny Arsenal" at Armed Forces Day observances. Maj. Gen. William K. Gormley, commander of the Ordnance Special Weapons Ammunition Command, and Col. Russell R. Klanderman, commanding officer of the arsenal, look on. (U.S. Army Photograph-Picatinny Arsenal.)

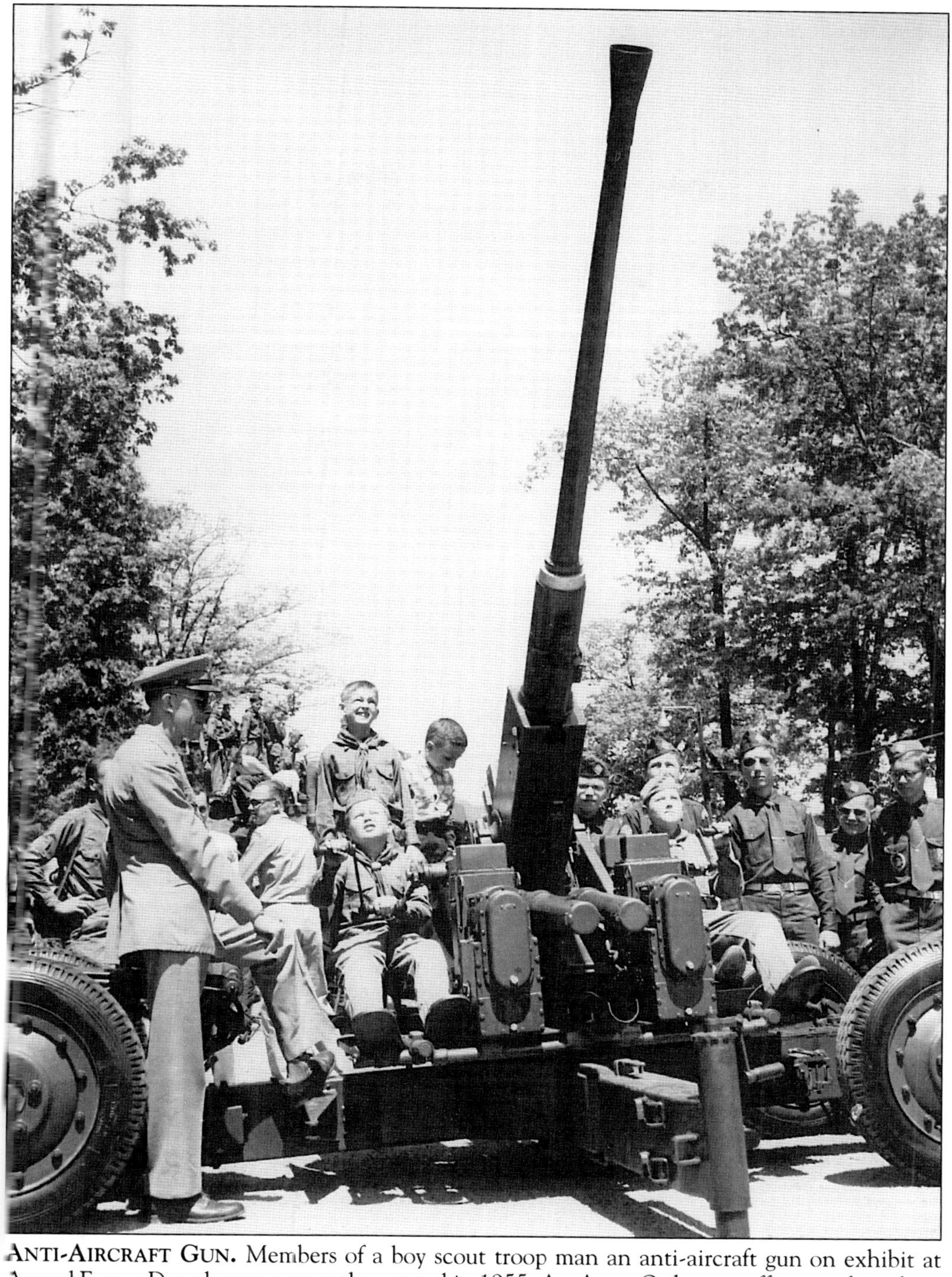

ANTI-AIRCRAFT GUN. Members of a boy scout troop man an anti-aircraft gun on exhibit at Armed Forces Day observances at the arsenal in 1955. An Army Ordnance officer explains how the gun works while other scouts look on. (U.S. Army Photograph-Picatinny Arsenal.)

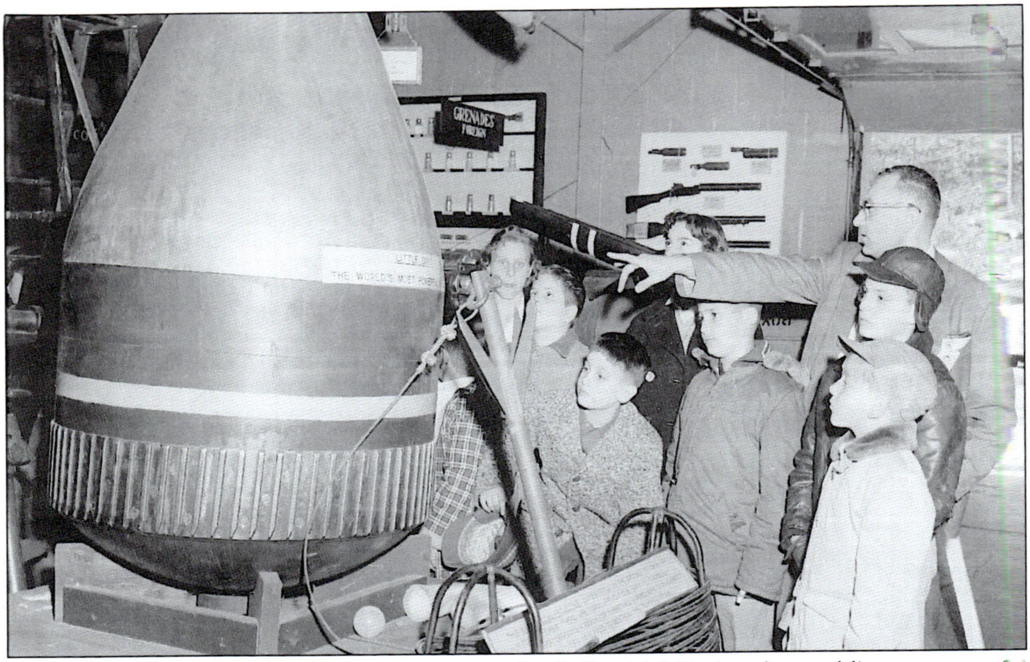

LITTLE DAVID MORTAR. Museum visitors view Little David, billed as the world's most powerful mortar. Developed during World War II in an attempt to get a bunker busting munition, it never got beyond the development stage. Its weight was 3,650 pounds, 1,589 of which were high explosives. It had a range of 7,000 yards and took two tractor trailers to tow it. (U.S. Army Photograph-Picatinny Arsenal.)

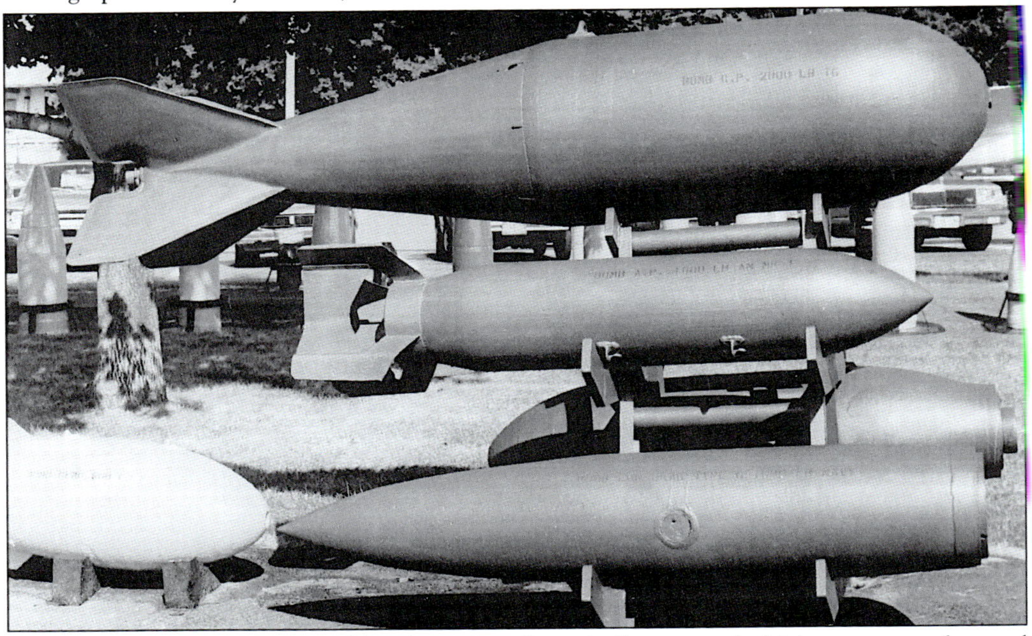

A BOMB DISPLAY. A display of bombs and artillery shells, some of which were manufactured and loaded at Picatinny, are visible in the arsenal museum park. Shown are 2,000-, 1,600-, and 600-pound bombs. In the background are artillery shells of various calibers. (U.S. Army Photograph-Picatinny Arsenal.)

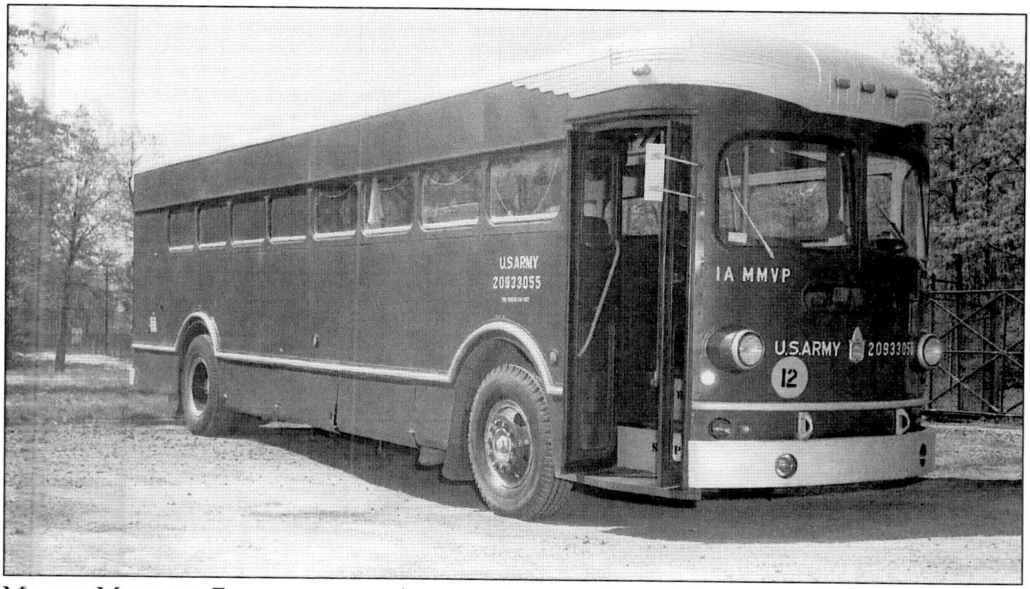

Mobile Museum. For many years, the arsenal operated a mobile van equipped as a traveling museum. It visited schools, museums, shopping centers, and special events throughout Northern New Jersey with its display of weapons, explained by an ordnance officer. (U.S. Army Photograph-Picatinny Arsenal.)

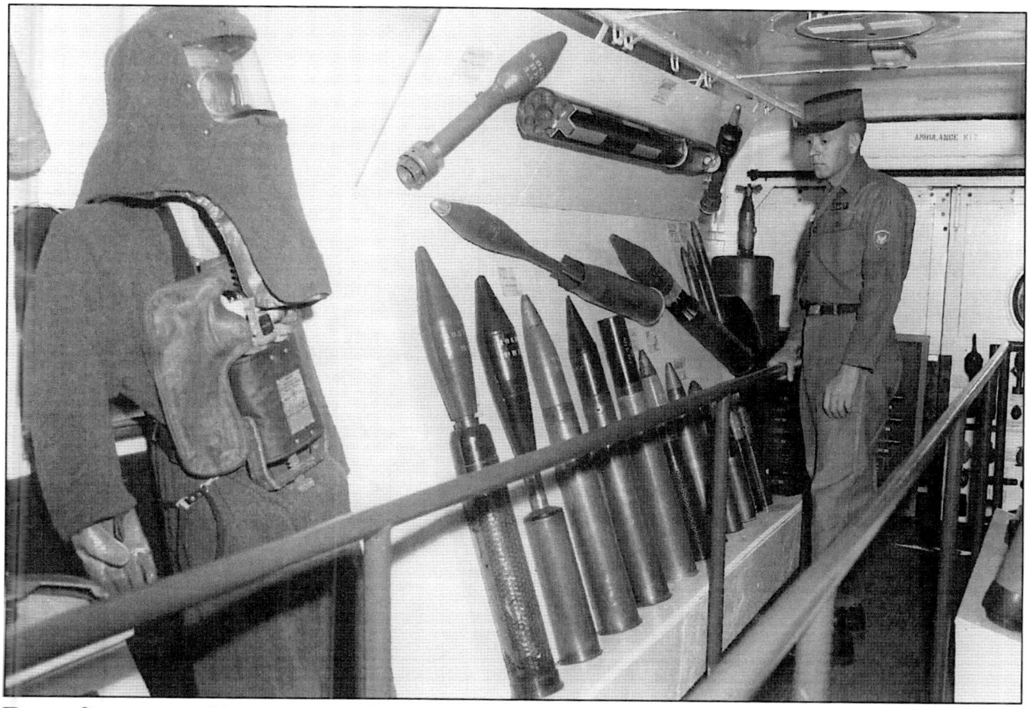

From Shells to Protective Gear. The interior of the traveling museum exhibited the major weapons on view in the museum, which was first established in 1920. Included were many munitions developed and manufactured at the arsenal. An ordnance officer explained when they were developed, special circumstances that required immediate development, and what some have been replaced by. (U.S. Army Photograph-Picatinny Arsenal.)

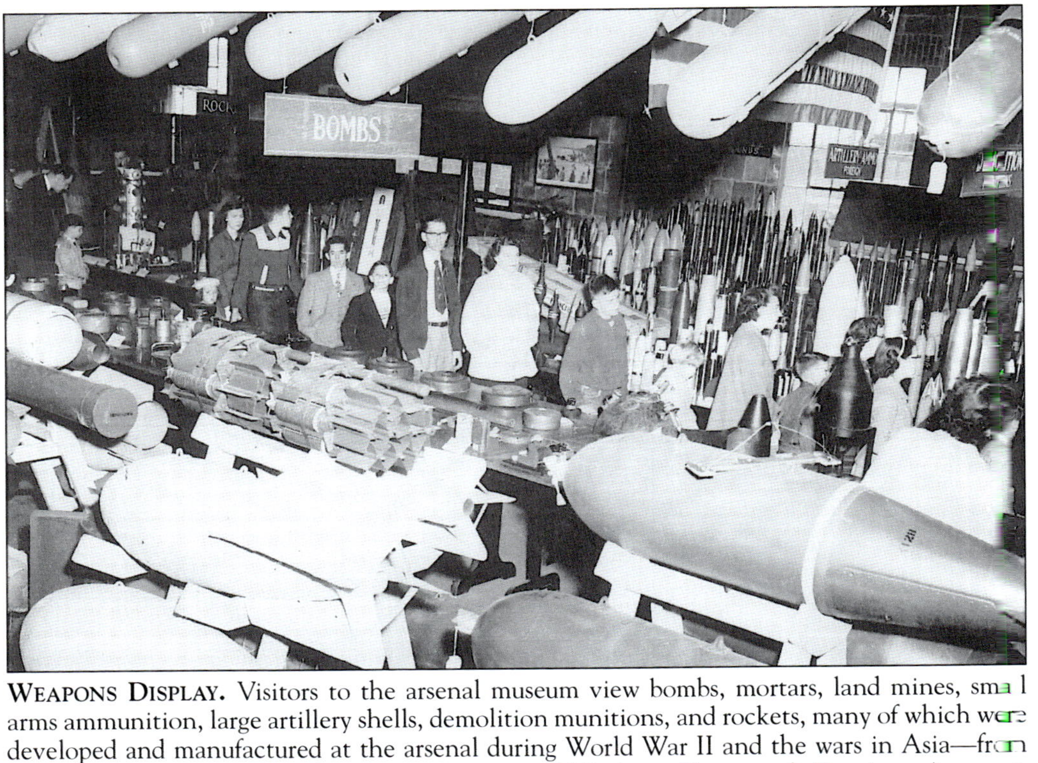

WEAPONS DISPLAY. Visitors to the arsenal museum view bombs, mortars, land mines, small arms ammunition, large artillery shells, demolition munitions, and rockets, many of which were developed and manufactured at the arsenal during World War II and the wars in Asia—from Korea to Vietnam. Pictures show many in action. (U.S. Army Photograph-Picatinny Arsenal.)

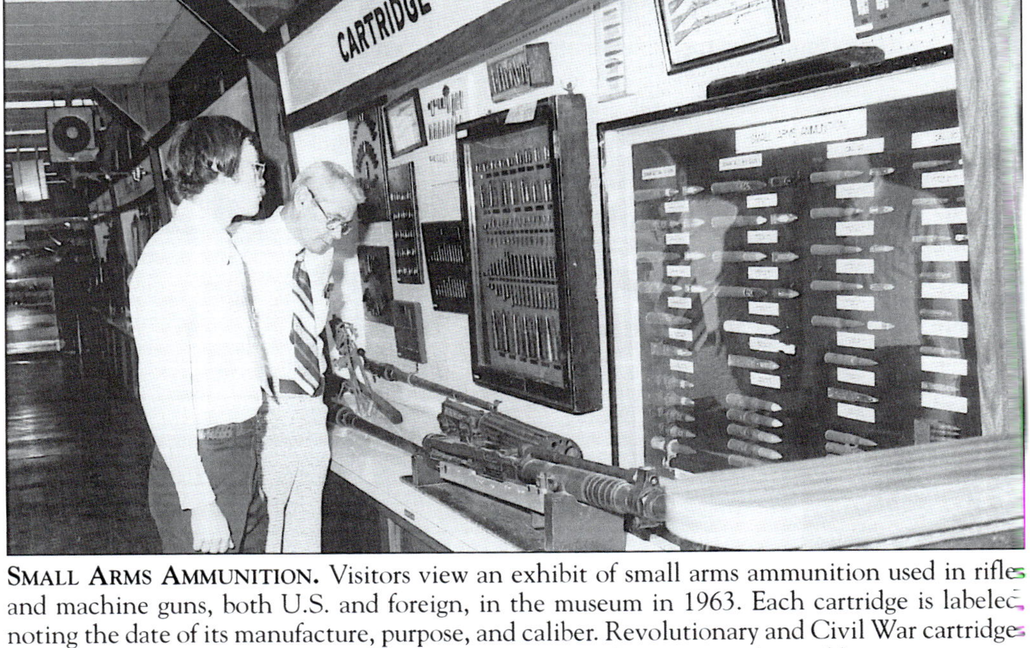

SMALL ARMS AMMUNITION. Visitors view an exhibit of small arms ammunition used in rifles and machine guns, both U.S. and foreign, in the museum in 1963. Each cartridge is labeled, noting the date of its manufacture, purpose, and caliber. Revolutionary and Civil War cartridges contrast with modern munitions. (U.S. Army Photograph-Picatinny Arsenal.)

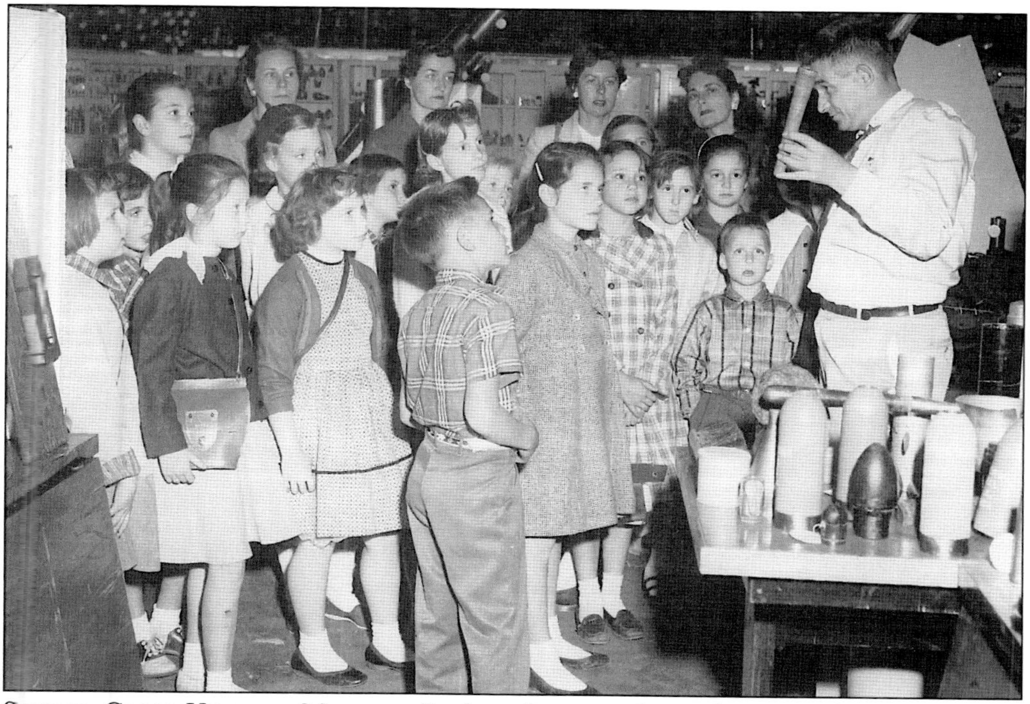

School Class Visit to Museum. Students listen to the explanation of various types of ammunition, their purpose, and detonation effects during one of many visits by North Jersey classes to the museum. The museum is open Tuesdays, Wednesdays, and Thursdays. (U.S. Army Photograph-Picatinny Arsenal.)

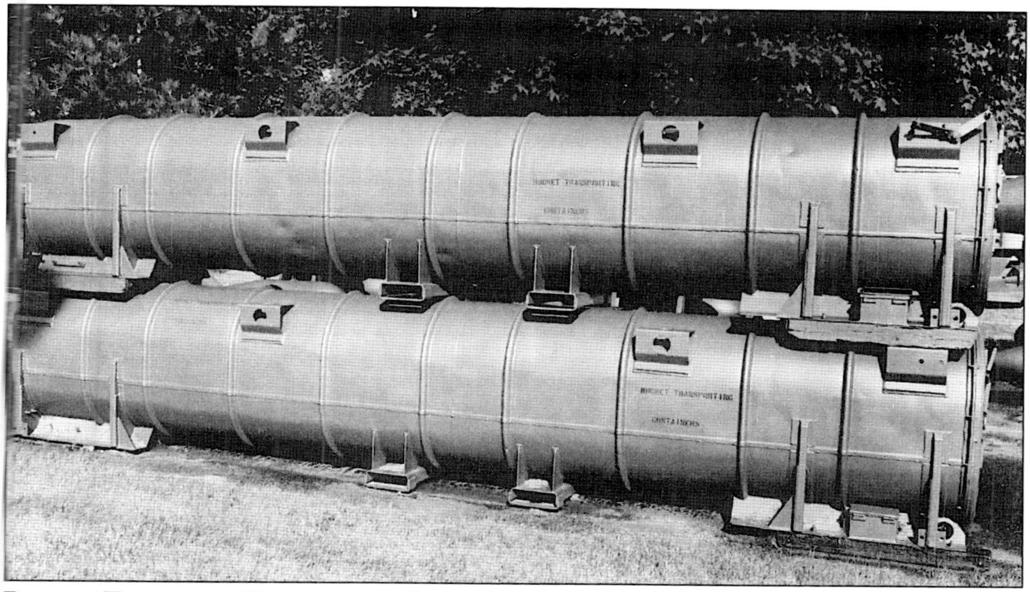

Rocket Transport Containers. Containers used to transport rockets are on display in the museum park. Manufacture of packaging at the arsenal dates to World War II, when in addition to producing ammunition, the arsenal had to get it to the front on time. Packaging engineers were imported to supplement the arsenal staff to develop better methods of packing bombs, explosives, and ammunition. (U.S. Army Photograph-Picatinny Arsenal.)

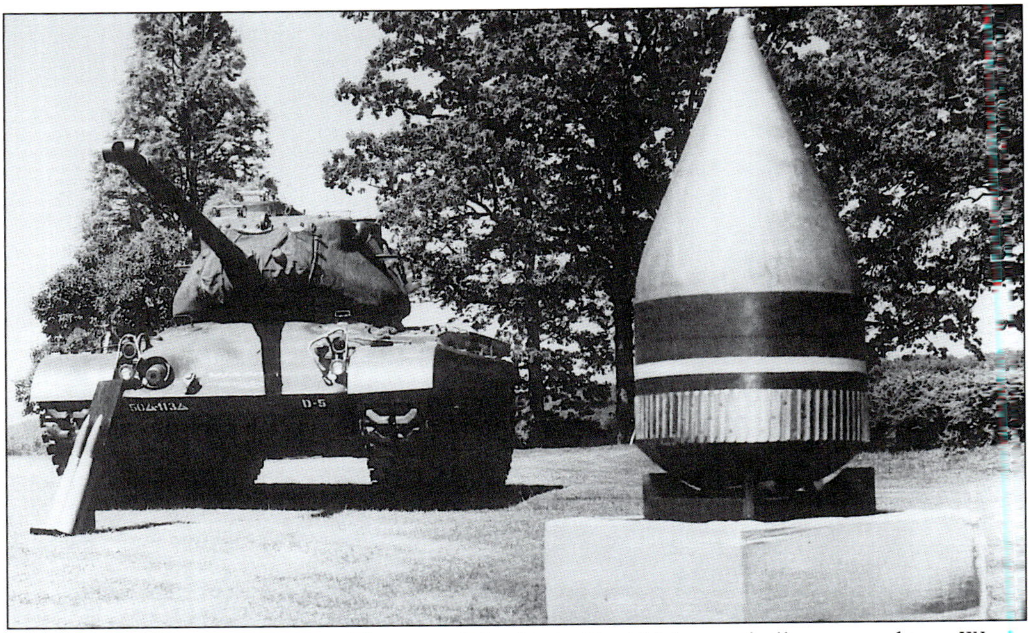

75TH ANNIVERSARY DISPLAY. The size of the Little David Mortar shell compared to a World War II tank is evident in this display at the 75th anniversary of the arsenal in 1955. Little David took a crew of five men to set it up and fire it. Artillery shells fired by the tank's cannon rest against the weapons front. (U.S. Army Photograph-Picatinny Arsenal.)

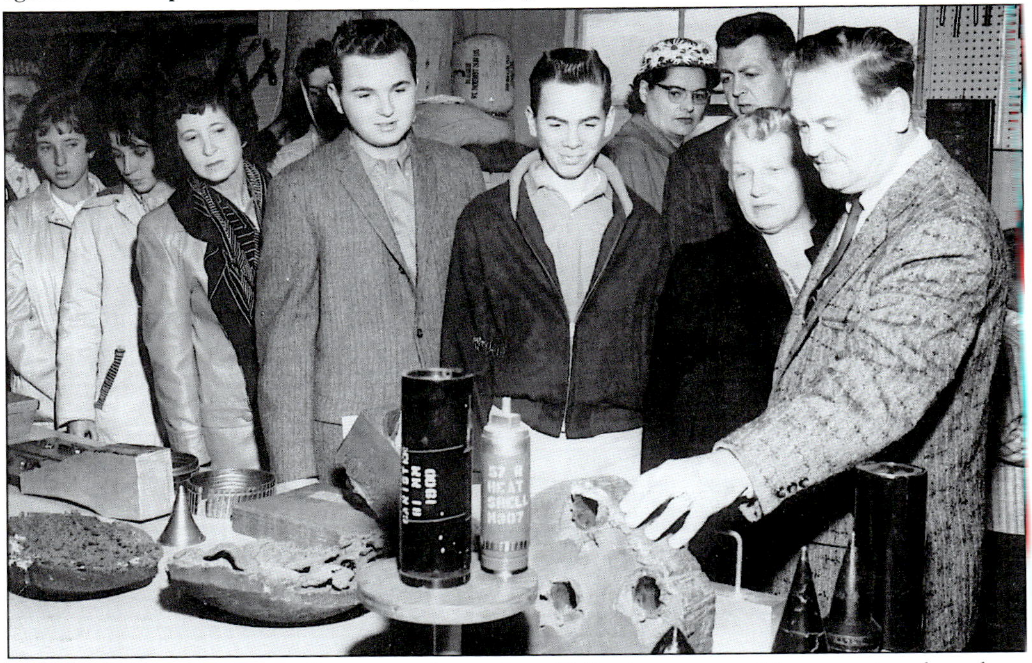

EXAMINATION. Visitors to Picatinny Arsenal in 1963 view ammunition, ranging from heat seeking artillery shells to small round shells, and see the damage they inflict. One item on display is an arsenal manufactured fuse discovered by occupation troops on the commander's desk in a munitions depot in Bavaria. It had been fired at German troops during the North Africa campaign in 1942. (U.S. Army Photograph-Picatinny Arsenal.)

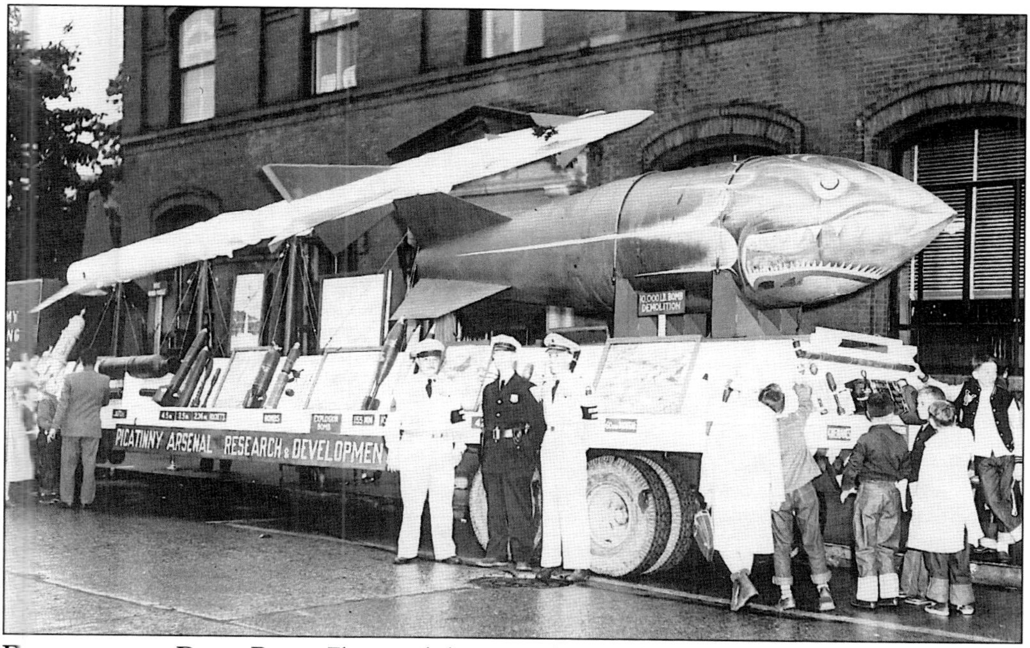

FLOATS HAD A DUAL ROLE. Floats exhibiting rockets and munitions developed at Picatinny had a dual role. Here a float is used in an Army recruiting drive in Washington, New Jersey, September 14, 1954. The shark-nosed bomb is a 10,000-pound demolition bomb, the first of which were loaded at the arsenal in 1942. (U.S. Army Photograph-Picatinny Arsenal.)

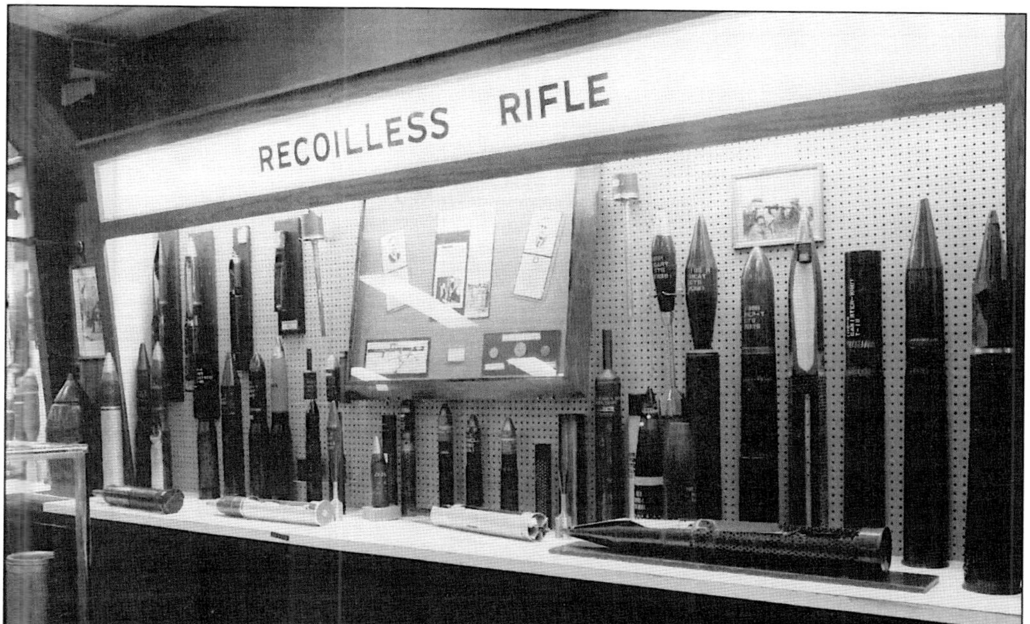

RECOILLESS RIFLE EXHIBIT. It was at Picatinny that ammunition for recoilless rifles, including high explosives, high explosive antitank, and smoke and canister rounds were developed and manufactured in the early years of World War II. This display in the museum shows the various caliber of shells manufactured plus a description of their use. (U.S. Army Photograph-Picatinny Arsenal.)

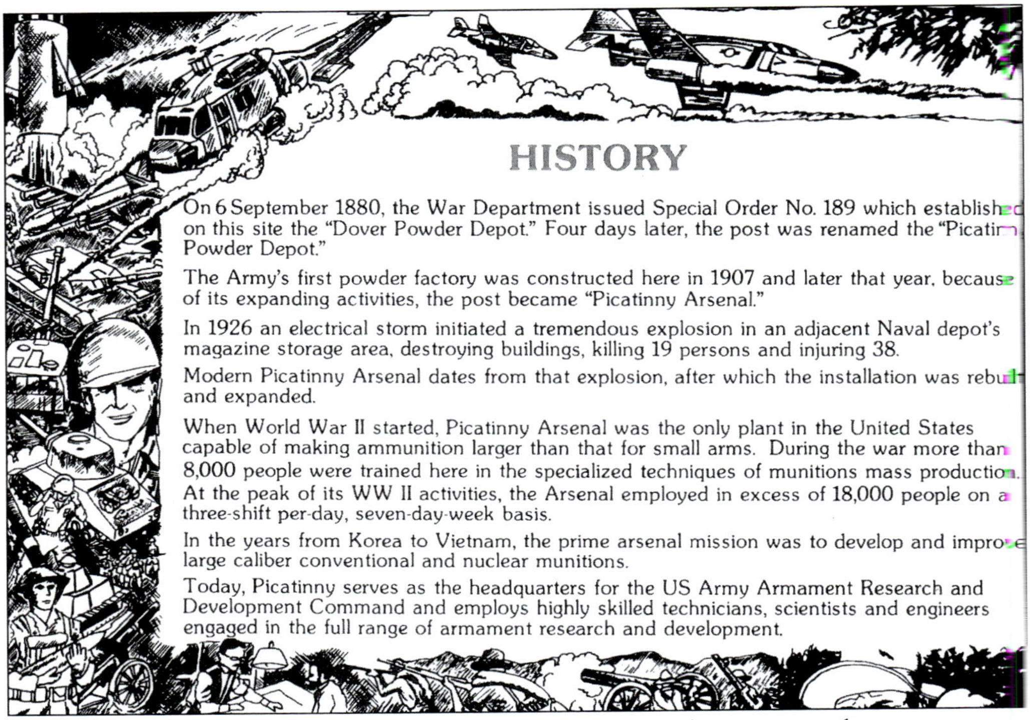

HISTORY

On 6 September 1880, the War Department issued Special Order No. 189 which established on this site the "Dover Powder Depot." Four days later, the post was renamed the "Picatinny Powder Depot."

The Army's first powder factory was constructed here in 1907 and later that year, because of its expanding activities, the post became "Picatinny Arsenal."

In 1926 an electrical storm initiated a tremendous explosion in an adjacent Naval depot's magazine storage area, destroying buildings, killing 19 persons and injuring 38.

Modern Picatinny Arsenal dates from that explosion, after which the installation was rebuilt and expanded.

When World War II started, Picatinny Arsenal was the only plant in the United States capable of making ammunition larger than that for small arms. During the war more than 8,000 people were trained here in the specialized techniques of munitions mass production. At the peak of its WW II activities, the Arsenal employed in excess of 18,000 people on a three-shift per-day, seven-day-week basis.

In the years from Korea to Vietnam, the prime arsenal mission was to develop and improve large caliber conventional and nuclear munitions.

Today, Picatinny serves as the headquarters for the US Army Armament Research and Development Command and employs highly skilled technicians, scientists and engineers engaged in the full range of armament research and development.

A DESCRIPTIVE LEAFLET. This leaflet explaining the history of Picatinny and its mission, past and present, was distributed to 10,000 visitors at Armed Forces Day observances May 15, 1982. It notes that the modern arsenal dates from the 1926 explosion, after which it was rebuilt and expanded. (U.S. Army Photograph-Picatinny Arsenal.)

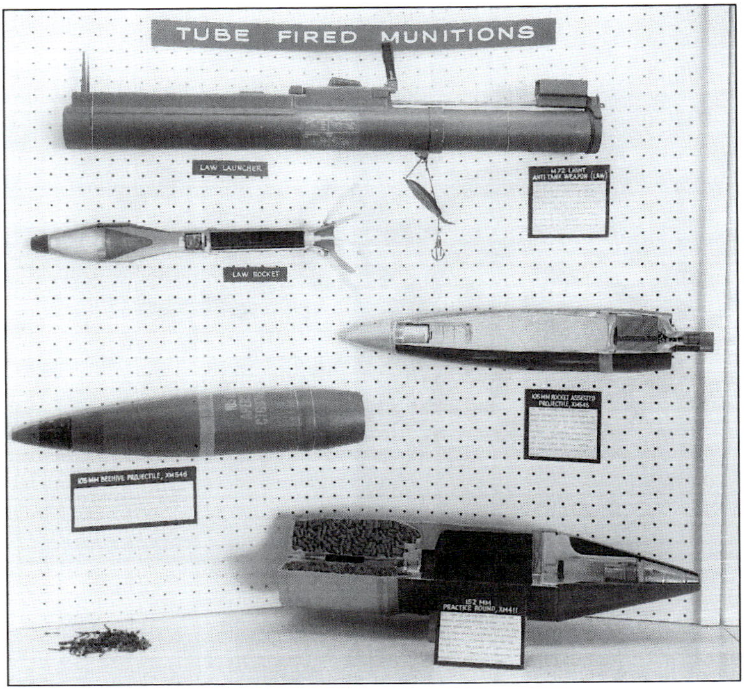

TUBE FIRED WEAPONS. This is a view of a museum exhibit of tube-fired, rocket-assisted weapons ranging from the Bazooka, the weapon that stopped Rommel's tanks in North Africa, to heavier projectiles. Included is the beehive shell, which was developed by arsenal technicians for use in the wars in Asia—from Korea to Vietnam, a 105-mm rocket assisted projectile, and a 152-mm practice round. (U.S. Army Photograph-Picatinny Arsenal.)